Professional Ethics in Construction and Engineering

Professional Ethics in Construction and Engineering

Jason Challender

WILEY Blackwell

This edition first published 2022
© 2022 John Wiley & Sons Ltd

The right of Jason Challender to be identified as the author of this work has been asserted in accordance with law.

Registered Offices
John Wiley & Sons, Inc., 111 River Street, Hoboken, NJ 07030, USA
John Wiley & Sons Ltd, The Atrium, Southern Gate, Chichester, West Sussex, PO19 8SQ, UK

Editorial Office
111 River Street, Hoboken, NJ 07030, USA

For details of our global editorial offices, customer services, and more information about Wiley products visit us at www.wiley.com.

Wiley also publishes its books in a variety of electronic formats and by print-on-demand. Some content that appears in standard print versions of this book may not be available in other formats.

Library of Congress Cataloging-in-Publication Data
Names: Challender, Jason, author.
Title: Professional ethics in construction and engineering / Jason
  Challender.
Description: Hoboken, NJ : John Wiley & Sons, 2022. | Includes
  bibliographical references and index.
Identifiers: LCCN 2021061395 (print) | LCCN 2021061396 (ebook) | ISBN
  9781119832096 (hardback) | ISBN 9781119828822 (epdf) | ISBN 9781119832102 (epub) |
  ISBN 9781119832119 (obook)
Subjects: LCSH: Construction industry--Moral and ethical aspects. |
  Construction industry--Employees--Professional ethics. | Building--Moral
  and ethical aspects.
Classification: LCC TH159 .C44 2022 (print) | LCC TH159 (ebook) |
  DDC 624--dc23/eng/20220125
LC record available at https://lccn.loc.gov/2021061395
LC ebook record available at https://lccn.loc.gov/2021061396

Cover Design: Wiley
Cover Image: © JU.STOCKER/Shutterstock

Set in 9.5/12.5pt STIXTwoText by Integra Software Services Pvt. Ltd, Pondicherry, India
Printed and bound by CPI Group (UK) Ltd, Croydon, CR0 4YY

C9781119832096_200722

For Tracy

Contents

Author Biography

Dr Jason Challender, MSc FRICS FAPM FHEA

Jason Challender has acquired 32-years 'client side' experience in the UK construction industry and procured numerous successful major construction programmes during this time. He is Director of Estates and Facilities at the University of Salford, member of its Senior Leadership Team and responsible for overseeing a large department of approximately 350 estates- and construction-related staff. Jason is also a Visiting Professor at the University of Bolton and an established author with 11 academic journals and 5 books published in recent years, all of which have been dedicated to his studies around the construction industry. Furthermore, he has previously participated as a book reviewer for Wiley. He has also attended many national construction and institutional conferences as a guest speaker over the years and is a Fellow and Regional Board Director of the Royal Institution of Chartered Surveyors and Board Director of the North West construction Hub.

Foreword

The subject of professional ethics has become increasingly topical and important in the construction and engineering industries over recent years on a national and global level. This has largely been predicated on high profile global cases that have come to light where unethical practices, standards and behaviours have been identified and widely publicised. These have involved breaches of governance and codes of conduct and aspects of corruption in more extreme cases, which has led to reputational damage for the organisations involved and a general loss of public confidence in the construction and engineering industries. Given that construction and engineering activities operate almost wholly in the public eye, it is inevitably scrutinised significantly more than other sectors of the economy, which present real challenges for those responsible for evidencing consistent, ethical practices across complex and sometimes challenging supply networks. Put simple, as a direct consequence of these inherent difficulties, I strongly believe that there is an absolute imperative for long term transformational change and the construction and engineering industries are at a critical juncture in this regard.

With his vast experience in the construction industry across 32 years, the author is very well placed to address the dilemmas and challenges around professional ethics and the book does provide a thorough articulation and analysis of the practical and pragmatic ways and means to approach the application of ethical standards. To contextualise the scale of the problem whilst offering best practice solutions, the author has covered many different areas including ethical dilemmas, regulation and governance, professional codes of conduct and corporate social responsibility. The book draws on case studies from the authors experience alongside hypothetical practical examples to describe certain ethical dilemmas and suggested best practices in many different scenarios. In addition, the author has also developed an ethics toolkit, as a practical guide for managing projects. It is hoped that this will encourage more ethical behaviours and standards and discourage breaches in immoral and illegal practices. The result is a professionally focused textbook aimed at vocational learners (at both undergraduate and postgraduate taught levels) and practitioners in construction, engineering, architecture, and the wider built environment.

This book is not the first to outline some of the ethical shortcomings in the construction and engineering industries, but I believe does offer what many have been waiting for. I hope this book offers the inspiration for more moral and responsible and ethical professional practices, in the overall pursuit of building more confidence and trust in the construction and engineering industries. In this sense, if it creates a new perspective of understanding and dialogue around professional ethics, then it will have served its purpose.

Dr Abdul Salam Darwish BSc MSCCITM PhD
PGCE DTQM DEPMP DMM DBM
School of Civil Engineering and Built Environment,
University of Bolton

Acknowledgement

The author would like to thank his family, Margaret, Kristin and Bobby for their encouragement during the book.

He would also like to acknowledge Dr Peter Farrell, Reader at the University of Bolton, whose long-term mentoring and has enabled the author to achieve all his career and academic successes. This book would not have been possible without his encouragement and proactive continued support.

List of Figures

List of Tables

List of Appendices

List of Abbreviations

ARCOM	Association of Researchers in Construction Management
AoC	Association of Colleges
BIM	Building Information Modelling
BIS	Department for Business, Innovation and Skills
CDM	Construction, Design and Management (Regulations)
CE	Constructing Excellence
CIAT	Chartered Institute of Architectural Technologists
CIC	Construction Industry Council
CIOB	Chartered Institute of Building
CMI	Chartered Management Institute
DfE	Department for Education
DV	Dependent Variable
EU	European Union
FE	Further Education
GDP	Gross Domestic Product
GOF	Goodness of Fit Indices
HE	Higher Education
HEFCE	Higher Education Funding Council for England
HESA	Higher Education Statistics Agency
HR	Human Relations
ICE	Institution of Civil Engineers
IV	Independent Variable
JCT	Joint Contracts Tribunal
KPIs	Key Performance Indicators
LCCI	London Chamber of Commerce and Industry
NAO	National Audit Office
NBS	National Building Contract
NEC3	New Engineering Contract
NHBC	National House Building Council
OJEU	Official Journal of the European Union
ONS	Office for National Statistics
PCFA	Principal Component Factor Analysis
RIBA	Royal Institute of British Architects
RICS	Royal Institution of Chartered Surveyors
SEM	Structural Equation Modelling

1

Introduction

Professional ethics in construction and engineering

> *Consider ethics, as well as religion as supplements to law in the government of man.*
> Thomas Jefferson (1807)

1.1 Introduction to the Book

The above powerful and unequivocal quotation from the former President of the USA reinforces the importance of ethics in the society and world we live in. It is perhaps the same quotation which has provided the focus for this book in an attempt to encourage construction practitioners to take a more proactive ethical stance in project management, change current working practices in the construction industry and improve project outcomes. Accordingly, the main focus of the book is to explore the role of ethics in construction management. In this regard, the overarching aim of the book is to create a factual client 'how to do it' guide or 'toolkit' for procuring more successful project outcomes. It is intended that this practical guide for construction practitioners can develop into a common due diligence framework on how to initiate, procure and manage construction projects and developments with ethical considerations at their heart. From this perspective, it will raise awareness of best practice and instil improvements in construction management with ethical compliance at the epicentre of project teams. It will seek to address the significant institutional risk that lies in the lack of a clear and consistent approach to ethics in projects and guidelines which are seldom universal and therein open to different interpretations. Such an approach will constitute a viable tool in ensuring effective, appropriate and successful interfaces of ethical standards and principles, via codes of conduct, in pursuit of improvements to construction management practices. Furthermore, it is also intended to provide an important insight into the influence of professional ethics in the success of construction projects and redevelopment programmes.

Despite the fact that few would disagree with the importance, values and principles of ethics, the practice of professional ethics has however traditionally not been an area for widespread reflection, consideration and focus within the construction industry. Cases have emerged over the last few years questioning the morality of the industry and the direction of its moral compass, bringing increased scrutiny upon it. In 2009 tender practices within the industry, associated with bid rigging or collusive tendering came under scrutiny by Competition and Markets Authority, resulting in multimillion-pound fines against over 100 main contracting organisations. *The 2017 Taylor Review* of modern working practices raised serious concerns about payment practices in the UK construction industry, especially the morality of supply chain management practices which have been seen to impose payment terms of up to 120 days onto sub-contracting organisations. In 2018, the collapse of Carillon led society to question the ethical practice of the industry. In addition, *Dame Judith Hackett's independent review of Building Regulations and Fire Safety in 2018*, following the Grenfell Tower disaster raised major questions about contractors' use of 'value engineering' to reduce costs as the expense of safety and once again led to society questioning the morality of the construction industry. Whilst events such as these continue to occur, bringing widespread negative publicity and reputational damage for the industry, rebuilding the confidence and trust of the general public remains a major priority. Farmer (2016) adopted medical metaphors, describing the construction industry as a sick and dying patient in urgent need of treatment if it was to survive.

Against this background, this book will provide further insight into the subject of professional ethics relevant to the construction industry and discuss the key areas of ethical standards, values and behaviours, environmental ethics, cultural change, governance and regulation amongst others. It will articulate, discuss and analyse some of the problems relating to adopting professional ethics and provide possible reasons to explain and understand why unethical practices exist. The potential importance of ethical principles and potential improvement measures to enhance and improve current practice, especially in the context of the UK construction industry, will also be articulated and discussed from different perspectives. Accordingly, in this regard the aim and objectives will be to expand the knowledge and breadth of understanding of professional ethics in the construction industry. Reflections will be presented together with possible recommendations for improvement measures and further research.

The book investigates the current arrangements that exist within the global construction industry, to create a more comprehensive understanding of the problems and dilemmas of unethical conduct. It explores and analyses the overall commitment of organisations and professional bodies to embed professional ethics into all construction stages which could be hindering the overall effectiveness of projects. This is intended to provide a suitable context for paradigm shifts in practice with measures to improve ethical values, principles and leadership as the catalyst for increasing project success.

A deficiency in appropriate and strong construction practitioner leadership to promote ethics in the construction industry has been highly documented by authoritative sources over many years (Boyd and Chinyio 2006; Kamara et al. 2002). The book will seek to address this ongoing ethical dilemma and act as a catalyst for improvements to the construction procurement processes. This is intended to encourage more transparency, honesty, integrity and openness between clients, their appointed consultants and the whole supply chain. This is a deliberate attempt to improve project management practices, which have arguably not been delivering the impact, and benefits in terms of successful project outcomes.

The book is intended to assist academics, construction-related practitioners and clients in their awareness, breadth of knowledge and comprehension of the issues around ethical considerations, with the overarching aim of delivering more successful project outcomes. This is felt to be particularly important as in previous studies into ethics in the built environment, very little attention has been focused on giving practical advice. The book has sought to infill the literature gaps through examination of traditional roles of clients, design consultants, main contractors and subcontractors and through providing guidance through the toolkit on potential improvement measures. Case studies and practical examples have been included to assist the reader on how theoretical perspectives can be applied to real-life construction projects and scenarios which involve ethical dilemmas. The book has also addressed academic calls for greater insight into how leadership around ethics can be created, mobilised and developed and more understanding of the resultant positive effects and impact that can be generated therein (Walker 2009). There will be frequent reference to construction practitioners' views and opinions throughout the book, and these have been sought through qualitative research carried out in 2020 from a small sample of semi-structured interviews.

There have been few books which have been written on the specific subject of incentivising appropriate professional ethics in construction specifically through a practical guide or 'how to do it' toolkit. Those which have been published have largely focused on theoretical studies examining different construction practitioner behaviours and relational analysis of clients with construction teams. Albeit the component elements of professional ethics have been covered previously and therein well-trodden ground, there has been very little to articulate how these can be incorporated into construction procurement strategies. The book, drawing on case studies from the authors' experiences and interviews, takes a different approach to professional ethics in the construction industry by asking some very fundamental questions:

- What is the importance and influence of leadership in influencing the strict compliance and adherence to ethical principles, values and standards?
- What is the extent to which good ethical standards, behaviours and practices can influence the success of construction projects?
- How can professional ethics be best embedded into the procurement of projects?
- What constitutes best practice and what is the extent to which the governance, regulation and practice of ethics can influence the success of construction projects?

In consideration of the above questions the book's objectives are:

- To be the standard reference for businesspeople in understanding how ethics affect projects and therein reducing construction-based risk.
- To explain in straightforward terms with practice-based examples, both real life and hypothetical, where ethical dilemmas may occur and how to steer construction practitioners to make the right business decisions and avoid immoral or illegal practices.
- To use case studies to look at patterns of ethical and unethical construction team behaviours and how these affect successes, failures and key risks on projects.
- To identify valued knowledge, skills, attitudes and behaviours and business practices that construction practitioners can use in their ethical approach to projects.
- To identify a set of clear guidelines, national or international, to support the adherence and regulation around ethical practices.
- To form the basis of a practical toolkit for guidance and teaching of ethics in the built environment.

The book is mainly intended for construction management practitioners and clients but could suit a wide target audience including under- and postgraduate students and academics. For its target audience it will provide a summary of best practice and guidance in upholding ethical values, principles and standards from initiation to post-handover across many different projects. For professional practitioners, the book will explore how ethical management can be a complimentary project skill and whether such skills can be proven to reduce risk in projects. Accordingly, it will provide the 'voice' of professional and institutional clients and present their recommendations and mechanisms for the creation of a common standardised ethical protocol. Furthermore, it will also consider whether this approach might compliment other developments in project management such as project team integrated working.

The books findings are presented to encourage professional practitioners to implement improvement measures through ethical governance mechanisms and initiatives. The introduction of such mechanisms is explained in the book and presented in a simple and effective way for improvements in construction project management practice. Reading this book will hopefully support the development of a deeper understanding of the benefits of having strong ethical leadership for improved outcomes for construction projects. With a better insight to how professional ethics can be instrumental to project success it should provide the potential to embrace the true philosophy of trust, confidence and how these can promote more improved construction management practices. The book is not intended as a holistic course textbook albeit it could be worthy inclusion on a recommended reading list for courses related to construction management. The toolkit for improving professional ethics for construction practitioners could be used as a basis for short-term training or conference proceedings for professional institutions and public sector organisations. Notwithstanding this, it is not intended solely as a practitioner guide. Rather, the book aims to cross this divide and provide useful insight to both academics and practitioners in developing their understanding of the topic area.

Understanding the risks posed for breaches of ethical standards by construction practitioners is a growing area as the tolerance for project failure reduces. Contemporary books to this seek to externalise professional ethics and codes of conduct by rooting them with professional institutions that create and regulate them. This book takes a different perspective and is unique in considering professional ethics as an integral and valuable component part of every construction stage. In doing so it creates a practical baseline for awareness of ethical considerations and therein smoothing the transition between traditional ethics teaching and embedding standard, values and principles in project practice. Accordingly, it is a starting point for standard ethical practices to be developed and integrated within project theory to support and de-risk subsequent project management.

The value proposition of this book is that it will be read, understood and accepted by businesspeople as their main guidance and reference tool for reducing construction-based risk associated with unethical conduct. Accordingly, this is a book written for businesspeople by businesspeople based on sound theory (how to do it) and sound practice (lessons derived from case studies). It is considered unique in that it represents a comprehensive and wide-ranging analysis of best ethical practice in the construction industry as well as other industries and sectors and what can be learnt from them. The book will take established and widely accepted business ethical management practices and models and align them with development/construction issues. From this perspective, it does not seek to adapt established project management systems and processes but simply use ethical best practice models to co-exist alongside them. The book also investigates the area of environmental ethics and corporate social responsibility. These have become more prominent and important over recent years with the advent of cases such as the Deepwater Horizon oil spill which had major financial implications and reputational damage for British Petroleum PLC.

Although the research was undertaken in the UK, and all findings are likely to therefore have best fit with the UK construction industry, the overall knowledge and understanding to be provided by this book will have international relevance. Other countries seeking to develop ethical guides using similar approaches to the UK will be able to utilise the book, with consideration of how the findings fit with their own understanding in practice.

Finally, it is worth acknowledging that the author has gained over 32 years' experience of construction management from both a practitioner and an academic perspective. From this, the book has drawn on both academia and practice, and it seeks from both these perspectives to prove an important insight into an area which has long been problematic for the construction industry.

1.2 Structure of the Book

The book has been structured into 12 chapters and covers a wide variety of different areas and consideration relevant to professional ethics in the construction industry, namely:

Chapter 1: Introduction
Introduction, justification, aims and objectives, readership and brief description of the areas covered throughout the book.

Chapter 2: Application of ethics in the context of the construction industry
Importance and significance of the construction industry, complexities around the bespoke nature of building projects and the risks associated with the dynamic and fragmented environment.

Chapter 3: The significance and relevance of ethics
Articulation of what ethics are, their origins and history, important to society and how to recognise unethical practices. Different roles and responsibilities of construction professionals and the ethical differentiation and non-alignment that this can create.

Chapter 4: Ethical dilemmas for construction practitioners
Explanation of how dilemmas that can be created, especially where commercial benefits are at stake offers practical guides and recommendations for maintaining ethical principles and practices under different scenarios. Non-adherence of codes of conduct and breaches of rules and regulatory standards and regulation and governance policies and procedures to avoid and address such adversities.

Chapter 5: Types and examples of unethical conduct and corruption

Identification of the various types of unethical and illegal practices from misuse of power to serious fraud. Articulation of what constitutes each type of unethical and illegal practice and how they can occur in a business context. Contains both hypothetical and real-life examples of situations and scenarios where breaches of ethical standards and illegal activity has ensued and measures that can be adopted to reduce unethical and illegal practices across the world.

Chapter 6: Regulation and governance of ethical standards and expectations

Focus on the importance of organisational transparency, regulation, and governance in the pursuit of ethical compliance, including management governance processes and procedures that can reduce the risk of unethical practices and behaviours.

Chapter 7: Ethical project controls in construction management

Articulation of what governance requirements can be instigated by construction professionals on their projects and presentation of some practical control mechanisms for regulatory adherence.

Chapter 8: Developing an ethics toolkit, as a practical guide for managing projects

A practitioner's toolkit designed specifically for the book, for encouraging best practice and to ensure that critical aspects of projects, including statutory compliance and quality controls are not compromised during the life of projects. In this way, it is intended as a practical guide for managing the processes and procedures linked to each aspect.

Chapter 9: Ethical selection and appointment processes for the construction industry

Highlighting the importance for construction professionals of the ethical processes around selection and appointment of their contracting partners to encapsulate the right level of transparency and sense of impartiality during tendering.

Chapter 10: Codes of conduct for professional ethics

Explanation of codes of conduct in regulating professional ethics, what these comprise and how they are applied to improve practices and behaviours. Articulation of the benefits of having codes of conduct and how they vary between different professional bodies and institutions. Importance of embedding codes of conduct, and ethical behaviours and standards into organisational culture by professional bodies.

Chapter 11: Implications in practice for ethics in the construction industry

Implications for the construction industry from unethical practices, including reputational damage and image considerations for organisations. Effects on construction project performance and brought about by unethical practices, especially focused on those developing countries around the world with construction quality data generated from research studies in Nigeria.

Chapter 12: Summary of key points, reflections, overview and closing remarks

Final chapter to summarise the book with reference to each chapter, extrapolating the key findings and issued raised. Presentation of the author's reflections and recommendations for the future of the construction industry, taking account of the inherent dilemmas for professional ethics.

1.3 Summary

The main focus of the book is to explore the role of ethics in construction management with the overarching aim to create a factual 'how to do it' practical guide or 'toolkit' to assist construction professionals procure more successful project outcomes. It will raise awareness of best practice and instil improvements in construction management with ethical compliance at the epicentre of project teams. It is also intended to provide an important insight into the influence of professional ethics in the success of construction projects and redevelopment programmes.

Even though few would disagree with the importance, values and principles of ethics, the practice of professional ethics has however traditionally not been an area for widespread reflection, consideration and focus within the construction industry. Cases have emerged over the last few years questioning the morality of the industry and the direction of its moral compass, bringing increased scrutiny upon it. High profile cases such as the Grenfell Tower disaster in 2017 and the collapse of Carillion PLC in 2018 have brought the construction industry into ill repute resulting causing reputational damage and loss of public confidence and trust in some cases. To respond to this issue, the book investigates the current arrangements that exist within the global construction industry, to create a more comprehensive understanding of the problems and dilemmas of unethical conduct. It explores and analyses the overall commitment of organisations and professional bodies to embed professional ethics into all construction stages which could be hindering the overall effectiveness of projects.

The book is intended to assist academics, construction-related practitioners and clients in their awareness, breadth of knowledge and comprehension of the issues around ethical considerations, with the overarching aim of delivering more successful project outcomes. Case studies and practical examples have been included to assist the reader on how theoretical perspectives can be applied to real-life construction projects and scenarios which involve ethical dilemmas. The book will hopefully support the development of a deeper understanding of the benefits of having strong ethical leadership for improved outcomes for construction projects. With a better insight to how professional ethics can be instrumental to project success it should provide the potential to embrace the true philosophy of trust, confidence and how these can promote more improved construction management practices.

References

Boyd, D. and Chinyio, E. (2006). *Understanding the Construction Client*. Oxford: Blackwell Publishing Ltd.

Farmer, M. (2016). *Modernise or Die: The Farmer Review of the UK Construction Labour Market*. London: Construction Leadership Council.

Kamara, J.M., Anumba, C.J., and Evbuomwan, N.F.O. (2002). *Capturing Client Requirements in Construction Projects*. London: Thomas Telford Publishing.

Walker, A. (2009). *Project Management in Construction*, 5th ed. Oxford: Blackwell Publishing Ltd. 150–158.

2

Application of Ethics in the Context of the Construction Industry

The construction industry is one of the most dynamic and complex environments as it is a project-based industry within which individual projects are usually built to clients' needs and specifications.

Tabassi et al. (2011)

2.1 Introduction

In considering the above quotation, this chapter of the book will articulate and discuss the importance and significance of the construction industry to global economies including the UK, whilst highlighting the complex nature of the sector. Such complexities around the bespoke nature of building projects will be covered alongside the temporary organisational aspects of the project procurement. Risks associated with the dynamic and fragmented environment that encapsulates the construction industry will be identified.

This chapter will also include the roles and responsibilities of different members of project teams including clients, project managers, architects, quantity surveyors, engineers, main contractors and subcontractors. It will consider these different roles from the perspective of ethical considerations and how they vary depending on the individuals. As a consequence of these differing perspectives, dilemmas around ethical decision making will be discussed. Ethical investigation models will be identified which seek to address such dilemmas in assisting construction practitioners in deciding what is right and wrong. Finally, the notion of building life cycles will be forensically analysed and the competing dilemmas of capital costs versus operational costs.

2.2 The Importance and Significance of the Global Construction Industry

The construction sector employs between 2% and 10% of the total workforce in most countries of the world (Rahman and Kumaraswamy 2005). In the UK, construction output in September 2020 accounted for more than £110 billion per annum and contributes 7% of Gross Domestic Product. Approximately a quarter of construction output is public sector and three-quarters is private sector. The construction industry is very

Professional Ethics in Construction and Engineering, First Edition. Jason Challender.
© 2022 John Wiley & Sons Ltd. Published 2022 by John Wiley & Sons Ltd.

diverse and includes elements ranging from the construction of buildings and infrastructure, maintenance, refurbishment to the manufacture and supply of building products and components. New build construction output accounts for approximately 60% of construction output whereas refurbishment and maintenance accounts for the remaining 40%. *Government Construction Strategy 2016–2020* outlines that there are three main sectors within the UK construction industry, and these are commercial and social (45%), residential (40%) and infrastructure (15%) (ONS 2020).

According to the *Government's Construction 2025 strategy* (HM Government 2013), the UK construction industry generates approximately three million jobs which accounts for approximately 10% of total employment and these include construction-related services and manufacturing. These jobs can be categorised as contracting which employs approximately two million people, and made up of 234,000 businesses and services, employing 580,00 people and 30,000 businesses. Furthermore, construction-related products accounts for 310,000 jobs and 18,000 businesses. The construction industry across the world can be regarded as a long-term industry linked to investment, albeit perceived to be a high-cost, high-risk industry by some. The performance of the industry can be seen to be a good indicator of the state and health of the sider economy. Accordingly, when an economy falters, it is common for the construction industry to slow down and when an economy begins to recover from a recession, the construction industry is one of the first industries to start to grow.

2.3 Ethical Challenges for the Built Environment

The issue of poor quality of construction projects and questionable professional ethics has been highlighted in the past especially in those developing countries. With the move over recent years globally to focus on quality-related outcomes rather than primarily cost and time, this has highlighted the need for construction professionals to focus on ethics.

Rahman et al. (2007) argued that the construction industry provides a 'perfect environment for ethical dilemmas, with its low-price mentality, fierce competition and paper-thin margins'. Transparency International which represents a global coalition against corruption found that 10% on the total global construction expenditure was lost to some form of corruption in 2004, which is staggering and further depletes the reputation of the industry. There have been growing reports over the years that unethical behaviours and practices are still in some areas of the construction industry are still common, and this is taking a toll on the industry's reputation. FMI undertook a survey which concluded that 63% of construction professionals felt that unethical conduct was still a problem for the industry. This clearly demonstrates that some improvement measures are required. Accordingly, the book will cover such ethical dilemmas and discuss measures, principles and models to curve such practices in the pursuit of a more regulated industry founded on moral values and behaviours. This is intended to act as a catalyst for improvements to construction management practices and improve the reputation of the industry which has suffered from reputational damage in the past. Later chapters of the book in this way will seek to equip practitioners with a practical guide or 'toolkit' to take business decisions which do not transgress into those areas of malpractice and unethical conduct. This is a deliberate attempt to promote best practice and regain the confidence and trust of the general public where reputations have been damaged over previous years. Hopefully this will deliver the benefit and impacts in terms of successful project outcomes and greater client satisfaction.

2.4   The Bespoke Nature of the Construction Industry

It is interesting to explore from where the construction industry has emerged as a bespoke project-based industry and there are many different theories around this. As a starting point, it is important to understand what makes the construction industry different and potentially 'at odds' from most other industries. In answering this question, it is worth contemplating that the procurement process around construction is very unlike that of most other industries. Those employed are made up of mostly small teams, ranging from construction workers to design consultants, who come together on a temporary basis for the life of a project and then disband to undertake different projects. This creates fragmentation and does not always allow the time for relationships to develop and flourish, which could be in itself a contributory challenge for trust generation. It is also important to reflect upon the 'end product' and that construction projects are nearly always bespoke to clients' requirements. This 'one off' or 'made to measure' aspect does, however, create risk and uncertainty for all parties. To fully appreciate and understand this context it may be useful to compare the procurement of a new building with the purchase of a new car. When one buys a car the make and model that suits your budget will be agreed alongside any affordable optional extras that are required. One can even 'test drive' the same model to ensure that it meets the customers' expectations in terms of feel and drivability. At this stage, on ordering the vehicle, the customer will know exactly what they will receive on the due delivery date, which is normally a few weeks at most and have an agreed fixed price. In this regard, there is very little risk that the customer will not receive exactly what they have expected when they ordered for the price they have secured. As the car is made in a factory it will be standardised, and quality control is normally very good accordingly. The complete opposite scenario could be argued to prevail when a building is being procured. It normally involves a prolonged period of time for design consultants to formulate a brief with clients, progress the design development and tender the projects to construction contractors. On receipt of tenders, this is where the process probably varies most from the car purchase example. In selecting the most appropriate tender it is important to consider the quality of the tenders rather than just accept the lowest price. Such factors as reputation, track record, resources and demonstration of an understanding of the project are vitally important as the quality, cost, scope and timescale in delivering the final project are normally anything but assured. There are three elements which are widely regarded as critical success factors namely cost, time and quality and these constitute what is commonly referred to as the 'iron triangle' which is illustrated in Figure 2.1.

There are several unknown factors in any construction process which could cause the cost of projects to increase, programmes to be delayed and the quality of builds to be

Figure 2.1 The iron triangle of cost, time, quality and scope.

compromised. This introduces two aspects which come into play around commerciality and risk and who incurs any additional costs is frequently an area where disputes arise between clients and their consultants and contractors. Furthermore, given that construction works normally incur significant amounts of money the stakes are high in terms of the final bill for clients and the level of profit attained by contractors. It is therefore perhaps not surprising, for reasons of commerciality, that parties to construction contracts have traditionally not relied on trust in dealings with each other especially around financial construction matters.

2.5   The Fragmented Nature of the Construction Industry

Despite the scale and importance of the UK construction industry, it is frequently criticised for performing poorly. In this regard it has been accused of being adversarial, wasteful and dominated by single disciples reluctant to innovate and poor at disseminating knowledge. It is also claimed that the industry is fragmented and according to Fewings (2009) there is a general tendency for those employed in the construction industry to ignore best practice owing to this fragmentation within the sector. In the UK this is reflected in the fact that nearly half of all work completed is undertaken by approximately 190,000 small contractors. Some of the workforce in this regard, especially those who undertake smaller jobs are often one-off individuals with limited and potentially insufficient experience. As a consequence of such inexperience there may be cases where their knowledge and awareness of issues relating to regulations, compliance, best practice and ethical considerations is lacking. In the context of the construction industry, which is associated with fragmented, complex and potentially confrontation practices, this can create a dilemma for the sector. It is, therefore, perhaps not surprising that there have been many reported cases in the press of examples of malpractice and contractors breaches regulatory standards and codes of practice. Some of these cases have resulted in litigation proceeding brought about by their clients which has resulted in reputational damage for the industry as a whole. In addition, these practices have frequently led to less than acceptable project outcomes in terms of value for money, delays and poor-quality build standards.

2.6 The Role of Construction Professionals in Managing Construction Projects

In the organisational structure associated with managing construction projects, construction professionals can be employed by clients directly as their employees and commonly referred to as 'client-side representatives'. Alternatively, they can be appointed by clients as consultants to manage the design and construction of building projects on their behalf. In this scenario they normally would form a client design team which could consist of consultant project managers, architects, quantity surveyors, mechanical and electrical engineers, structural engineers and other specialist consultants (Inuwa et al. 2015). It is normally the role of project managers and the overall design team appointed and led by clients who are responsible for developing the requirements of their clients, setting project briefs and managing the overall construction processes. Clients also employ their main contractors directly, and the main contractors under a traditionally procured building projects will employ subcontractors. It is normally the subcontractors who appoint their respective suppliers and specialist sub-subcontractors.

The various aforementioned roles and responsibilities of construction professionals are detailed below.

2.6.1 Construction Clients

Clients normally represent the sponsors of the project and therein have overall responsibility for the cost of the construction works and most of the risks involved. As they will be the 'end users' of buildings that are procured it is important for them to ensure that buildings are designed and constructed to meet their detailed and prescriptive brief and functional performance requirements. In this regard, clients need to determine and set down the goals for their projects and the means by which they intend to deliver and achieve them. In this pursuit it is vital that they appoint the right specialist consultants who are experienced and offer expertise for the nature of a particular project. It is also imperative that they ensure that the right balance of resources is deployed in other areas, and this includes budgetary and time considerations to deliver projects safely and to an acceptable quality. For this reason, clients need to be equipped with the tools to lead their projects which calls for informed and methodical approaches with clear responsibilities, roles and decision making.

Having clients at the forefront of projects with all the necessary skills in leadership, knowledge and resources has been highlighted as one of the main ingredients in improving the overall performance of the construction industry (Challender and Whittaker 2019). Furthermore, maintaining the 'client voice' in decision making and identifying the brief have become more important in recent years and not simply delegating these responsibilities to others in the professional team. This proactive leadership in the construction process, albeit challenging especially for 'lay construction clients', is essential in avoiding poor construction outcomes associated with time, cost and quality implications. In recent years the calls for transformational change through the Farmer Review of the UK Construction Labour Market (Farmer 2016) and the Construction Sector Deal recognise the vital role of clients in striving to 'turn the tide' and spearhead such changes. Despite this, the construction industry has been slow to embraced strategies linked to client leadership. Accordingly, the unique role of clients and their preparedness in projects is emerging as a 'hot' topic. This is not sufficiently covered by the professional institutions where the focus is on development of industry professionals rather than clients. Previous research within construction projects has mainly revolved around the development of professional teams which is well-trodden ground and has increasingly diminishing returns on risk reduction.

The construction industry is a risk reward venture undertaken by clients alongside their appointed project teams. Notwithstanding this premise, it is clients who ultimately take the biggest risk and if they represent a typical small business, arguable possess the least knowledge of protocols and culture on the construction industry (Challender and Whittaker 2019). Within the structure of a construction project, a series of government-sponsored reports *Constructing the Team* (Latham 1994), *Rethinking Construction* (Egan 1998) and *Accelerating Change* (Egan 2002) have made radical changes to the construction industry, making it more client focused than ever before. A greater sense of team working and integration between clients and the design and construction supply chain has now significantly reduced design and construction risk to, and from, the client. However, through the lens of designers and constructors the client can still be seen as a 'risk'.

Part of this risk can be characterised as the dynamic shift that a novice client has to undertake to become a '*Developer*'. The client is required not only to take on the role of delivering their own business and operational change but also choose the right resources and create the right environment to successfully and seamlessly deliver a construction project. This is an enormous risk and time consuming to the corporate business putting great strain on its management and physical resources. For many it is a 'leap into the dark' – the construction industry having few parallels in manufacturing.

2.6.2 Construction Project Managers

These construction professionals are normally regarded as having the overall responsibility for projects and accordingly are referred to as the lead consultant. They are responsible for all stages of projects from conception through to completion and sometimes involved in the 'in use' occupation phase also. Their role includes allocating resources for projects and managing the other design team consultants to ensure that successful outcomes are achieved. This is especially important considering the time, cost and quality aspects of projects. Accordingly, construction project manager responsibilities are arguably what guides a project to succeed. Construction project managers tend to have both good technical skills and leadership skills in directing their teams. In the latter case, communication skills are paramount in motivating a large project team and coordinating different contributions from many individuals.

Construction project managers work closely with architects, engineers and quantity surveyors to develop plans, establish programmes with timelines and milestones and determine budgets for buildings. In developing these plans, they are ultimately responsible for ensuring that projects are completed on time, on budget and meet all requirements of the client brief. Their role also includes reporting to their clients on the progress of projects at different stages and manage relationships with key stakeholders and external bodies.

2.6.3 Architects

These construction professionals are trained and qualified in the science and art of building design and normally in the UK members of the Royal Institution of British Architects (RIBA). Their role is to develop the concepts for buildings and structures and transpose these concepts into plans and images. Architects are responsible for creating the overall aesthetic and appearance of buildings and other structures whilst ensuring that they are correctly designed to perform well against a predetermined client brief and specification. In this regard, they must ensure that buildings are functional, safe and economical and meet the requirements of the end users. They need to consider how buildings will be used and the types of activity that will be conducted within them, whilst meeting all statutory requirements including building and fire regulations.

Architects provide various designs and present drawings, illustrations and a report to their clients based on the project brief. Computer-aided design (CAD) and Building Information Modelling (BIM) technology has for many years replaced traditional drawing as the most common method for creating design and construction drawings. The role of the architect is not completed at the design stages but continues on through the construction phases wherein they have a responsibility to oversee and supervise the works. In the construction phases they are frequently requested to enact changes to the drawings and specifications in response to their client's changing requirements. Other variations may be predicated upon unforeseen site circumstances and physical constraints not envisaged at the design stages. These may include budgetary considerations or unforeseen ground conditions. As construction proceeds, architects must ensure that the construction contractors adhere to the agreed designs and specifications and use the correct materials and products. They are also expected to check and monitor adherence of the ongoing works from a construction workmanship-quality perspective and ensure that the project programme is maintained.

2.6.4 Structural Engineers

Structural engineers in the UK are normally members of the Institution of Structural Engineers. They design and oversee all structural aspects of building and infrastructure projects. This includes undertaking structural calculations, considering proposed loading and forces on building elements. This would involve them in the preparation of structural design drawings and reports to ensure that buildings are completed to a safe and compliant standard and avoid damage or collapse when loaded. Alongside buildings, other structures can include retaining walls, bridges, foundations, structural frames and roads.

Structural engineering is related to the research, planning, design, construction, inspection, monitoring, maintenance, refurbishment and demolition of permanent and temporary structures. It also considers the economical, technical, aesthetical, environmental and social aspects of structures. On progression of the architectural design, structural engineers are required to design the main structural elements of buildings which include the foundations, walls, floors and walls. Structural engineers make decisions on the type and quality of materials used for structural elements and this could include designing concrete to meet the correct loading requirements or strength. In this regard they determine the thickness, span and depth of structural members in the build such as beams and the size, quality, quantity of type of concrete reinforcements.

Structural engineers' designs are based on codes which vary with countries and regions of the world and normally regulated by national building regulations. Such regulations are normally predicated on the premise that under the worst load the structures will be subjected to in their lifespans, they must remain safe. During the construction phase structural engineers regularly inspect materials and supervise the construction of structural members. They are required to approve completed works and this may sometimes involve testing materials. An example could be cube crushing tests to ascertain a required concrete strength has been achieved. In the event of building failures or collapse, they are brought in to carry out investigations and determine the causes, effects and solutions against reoccurrence.

2.6.5 Quantity Surveyors

A quantity surveyor is a construction industry professional who specialises in estimating the value of construction works and managing all aspects of cost control on projects. They are normally members of The Royal Institution of Chartered Surveyors (RICS) and projects they are involved with could include new buildings, renovations or maintenance work. Furthermore, they can be employed on a wide variety of projects covering all aspects of construction such as civil, mining and infrastructure projects to determine the cost of such facilities. They are normally responsible for calculating early design costs for budgetary purposes through composite rates. When more design information becomes available, they can provide more accurate budget estimates using elemental cost methodologies. Quantity surveyors work closely with the other design team members and assist in the tender process. They produce tender documents including detailed pricing schedules known as a bill of quantities and manage the tender process, evaluating bid submissions alongside the other design team consultants.

The term 'quantity surveyor' is derived from the role taken in quantifying the various resources that it takes to construct a given project, such as labour, supervision, plant and materials. They estimate and analyse the effects of design changes on budgets and deal with most of the contractual issues that arise on projects, including claims for variations and disputes. This involves overseeing elements of contract administration, clarifying and evaluating tenders, controlling variations, assessing contractual claims and negotiating and agreeing final accounts. They also are responsible for valuing completed work and arranging for payments through a certificate of payment contractual process and sometimes act as expert witnesses in litigation cases.

2.6.6   Building Service Engineers

Building services engineers are responsible for designing and overseeing all the mechanical and electrical building services installations on construction projects including lighting, heating, gas installations and all electrical systems. They provide all the building services to give buildings functionality, and which provide a stable internal environment to ensure that they have the correct temperature, air quality and lighting levels. This requires the provision of all the necessary backup support systems such as power, hot and cold water and lifts. In addition, the installation of life protection systems such as fire alarms, and sprinkler systems is an important responsibility for building services engineers. These functions should ideally be linked to sophisticated building management systems to ensure effective control and to minimise energy consumption.

In the UK they are normally members of the Chartered Institution of Building Services Engineers (CIBSE). Building services engineers work very closely with the other design team consultants to ensure that their mechanical and electrical proposals are aligned and coordinated with the architectural and structural designs for a given building. They influence the architecture of a building and play a significant role on the sustainability and energy demands of building. Managing energy has become more important in recent years with the need to reduce carbon generation and buildings to be more environmentally sustainable through use of renewable energy. Finally building services engineers oversee the testing and certification of mechanical and electrical installation towards the completion of projects to ensure that they meet regulatory requirements, comply with the agreed performance specifications and conform to their designs. This is to safeguard that building services installations are fit for purpose and have not been compromised by the mechanical and electrical specialist subcontractors.

2.6.7   Main Contractors

A main contractor, sometimes referred to as a general contractor, or prime contractor is responsible for the day-to-day oversight of a construction site, management of vendors and trades, and the communication of information to all involved parties throughout the course of a building project.

It is the responsibility of the main contractor to execute all construction work activities that are required for completion of projects. Their role takes on board many different aspects of construction management including project planning, managing all health and safety aspects, overseeing legal issues, coordinating and supervising all construction activities and undertaking contract administration duties. They also appoint and manage all construction subcontractors and ensure that the sequencing of all subcontract packages of work activities are closely and carefully coordinated. In this regard, the main contractor brings a team of all the required professionals together, overseeing the construction whilst ensuring that all necessary measures are taken to execute a project effectively.

The main contractor is responsible for preparation of plans and programmes to carry out construction projects. This ranges from hiring construction workers to developing a step-by-step timeline that the project will follow from start to finish. They are also responsible for hiring, supervising and payment of subcontractors and suppliers alongside obtaining materials for the project to meet specifications. Their other duties include acquiring all the necessary licenses and permits from the local authority or relevant body so that building projects can progress legally and safely. Table 2.1 includes many other areas of role and responsibilities that main contractors undertake in the course of manging construction projects.

Table 2.1 The role and responsibilities of main contractors.

Project Planning Responsibilities

Planning around important project development and implementation stages in advance of construction work commencing

Managing various issues on projects such as the requirement for materials and equipment

Anticipating any potential modifications on projects

Safeguarding that health and safety specifications are followed at all times

Practicing excellent communication between all parties involved in the construction process such as clients and subcontractors

Determining and addressing legal and regulatory requirements

Project Management

Managing the budget for the completion of construction activities

Responsibility to find and appoint the right subcontractors and individual suitable for the nature of a given project

Managing relationships with subcontractors and coordinating the equipment, materials and other services required by them for the duration of the project

Submitting detailed valuations for payment of works completed

Project Monitoring

Monitoring of projects in terms of project programmes and key milestones

Monitoring the quality of work carried out and adherence to detailed specifications and specialist design drawings

Reviewing, modifying and updating project programmes and risk assessments to reflect the current status of projects, taking account of any variations to the scope of the works

Implementing buildability measures throughout construction contracts, deploying value engineering methodologies and therein offering cost-effective building solutions

Legal and Regulatory Responsibilities

Obtaining all necessary building permissions and permits

Ensuring that projects are carried out in compliance with legal and regulatory issues

Health and Safety Responsibilities

Monitoring all health- and safety-related issues

Creating viable and workable safety policies to ensure health and safety is practiced and always maintained. This may entail risk management strategies, emergency response systems and other preventative measures to ensure site safety

Ensuring that all construction site staff use safety equipment in the course of their work and comply with method statements and risk assessments

Providing safety awareness for all staff

2.6.8 Subcontractors

According to NRM1 (RICS 2012) the term 'subcontractor' means 'a contractor who undertakes specific work within a construction project'. Normally subcontractors are specialists in a particular area or trade. They will perform all or part of the obligations of the main contractor's contract. In this regard, subcontractors have similar responsibilities as main contractors in that they must adhere to policies and procedures on site, provide personal protective equipment for their construction staff, ensure that the equipment they use is safe, report safety hazards and incidents. They need to have regular communication with the main contractor and report to the main contractor's project manager on the progress with their work package. Any issues or problems in their work area must be communicated to the main contractor without delay to avoid any knock-on delay implications on the overall construction programme. This is largely owing to the dependency that one subcontract work package can have on another owing to the sequential nature of construction work. This is especially important when a work package is on a 'critical path' and any delay to its completion will prevent another work package commencing.

Normally a large construction project or renovation often involves many different subcontractors who work together to complete projects in a timely manner. The nature of the subcontract works will determine the timing and sequence of when it is undertaken on site. For instance, a groundwork or piling subcontractor will often be one of the first subcontractors to commence and complete their works. Conversely a decorating or flooring contractor may be one of the last subcontractors in the sequence, as their work is planned at the end of the contract programme. Depending on the subcontract agreement they may be responsible for providing their own materials and equipment for the work package there are employed to undertake.

2.7 Different Perspectives on Ethics

The meaning of ethics can vary depending between the different roles and responsibilities of many different individuals that make up project teams. For instance, ethics as viewed from a client's perspective could revolve around a project team which is honest, truthful and presents accurate and transparent progress reports and information on at all times. Having a project team that can be trusted is therefore of paramount important for clients who will ultimately be the sponsors of the project and responsible for the cost of the construction works. Accordingly, building relationships with the team from the client's perspective is heavily dependent on having good communications and trust in those they employ. Clients will also expect ethical behaviours from their teams in managing projects closely and reporting any potential risks. Clients need to be made aware of any potential issues on projects especially related to cost, time and quality matters. If their project teams do not keep them fully appraised of issues that can potentially cause them difficulties, then this could be regarded as unethical behaviour and likely to lead to a loss of trust. In situations where unexpected incidents on projects arise which have not been predicted by those responsible for managing, this can lead to a loss of confidence from the client's perspective. Similarly, if members of the project team are perceived to charge clients monies which are deemed to be unreasonable and not what they agreed or expected then once again they could result in a loss of confidence. Another common source of frustration for clients can emanate from contractors not dealing with defects and snagging post contractor. There is a contractual and ethical responsibility for contractors to address reported building issues in a responsible and timely way, and delays which can cause clients disruption can cause tensions in client–contractor relationships.

From the consultant project teams perspective, they may be other ethical considerations to those of their clients. These may include expectations that clients are transparent in keeping them abreast of all issues especially around budgets and any issues that may affect progress of construction projects. For instance, if clients require approval from third-party stakeholders, e.g., funders, then some of the project team should be aware of dates in this regard so that these milestones can be accommodated into the programme timelines. The other ethical element from the project consultants' perspectives is to be reimbursed for any additional work which they are instructed to undertake. What can be sometime frustrating and damaging for client–consultant relationships, according to Fewings (2009) is where clients as paymasters expect their consultant to undertake additional scope of services at no cost.

From main contractors', subcontractors' and suppliers' perspectives, their view on ethics can be different again from clients and consultants. One of their main concerns is to be paid on time. Timely payments for these parties are especially important as cash flow for them is critical in their financial dealings. A lot of construction companies are forced into administration as a direct result in delayed payments to them. Accordingly, contractors and suppliers are sensitive to any instances where their employers pay them outside timescales which they have contracted to. This practice is regarded as unethical and could lead to disputes and in some cases legal actions being pursued.

In addressing some of these ethical issues from the clients', consultants' and clients' perspectives, Challender (2017) articulated that building collaborative working relationships between the project team including the wider supply chain is essential. A 'collaboration toolkit' was devised by in this way, with intervention measure to promote trusting relationships and partnership working. This was designed to manage problematic issues on projects in a proactive and responsive way before they progress to potential disputes between parties.

2.8 Decision Making from an Ethical Standpoint

The construction industry, as previously articulated, is dominated by many different organisations and individuals from a whole range of diverse backgrounds employed as part of their roles to make decisions of behalf of others. There is an implied duty of care alongside established values and standards which govern whether decision making can be deemed good, bad or indifferent. This can help inform whether decisions are right or wrong in a situation or context. In theory it would be easy to surmise that it is clear and obvious in most cases what represents poor decision making. However, in practice the picture is not always black and white but different shade of grey brought about by

Table 2.2 A model for ethical investigation.

Ethical Category of Investigation	Description
Descriptive ethics	Utilises experiential learning and empirical studies to aid and direct comprehension within a certain context. Gives precedents and principles for guidance for decision making.
Normative ethics	May provide one answer dependant on a particular view that is adhered to.
Philosophical ethics	Provides a moral answer by defining the dilemma but in a very theoretical way.
Practical ethics	Provides a 'moral compass' as a guide to acceptability. Provides practical models and rules of application to suit organisational values.

subjectivity rather than objectivity. This can create problems for the construction industry in deciding how we judge what is right and wrong and therein what is acceptable. To assist in this dilemma, Harrison (2005) created a model for ethical investigation predicated on different branches of ethics. These are summarised in Table 2.2 and can assist in deciding whether a particular situation or decision is ethical or unethical.

In consideration of the above, more than one approach may be used and really depends on the context to the situation being investigated.

According to Fewings (2009) there are different models relating to ethics in decision making. The 'moral intensity model' is related to the moral impact on decisions by reflecting on the extent of the consequences that may ensue from certain decisions. It would also seek to assess and consider the social backlash from certain options and the probability that negative outcomes could result from those decisions. In such circumstances it would try to predict which stakeholders or members of the community could be affected and to what degree. Some would argue that considering only the negative outcome of decisions is unethical in itself as could be construed to be a 'damage limitation' exercise and not derived from ethical and moral judgements to do the right things and adopt the most ethical pathway.

The 'business model' (Fewings 2009) is a commercially driven approach based on maximising financial outcomes and values. This is predicated on maximising profits but considering whether approaches are ethical and legal in this process. For this reason, it is important for an organisation utilising this model that the legal and ethical considerations are not outweighed by maximization of profits. We have seen many different examples in the past especially in financial services sector where financial considerations have dominated the *modus operandi* of companies and where a culture of non-compliance to legal and ethical standards has emerged. The global recession of 2009 was allegedly brought about irresponsible and unethical merchant bankers and the downfall of old established banks such as Merrill Lynch or Wall Street. There are many lessons to be learnt from such cases and with this approach it is important to have closely managed supervised robust procedures for adherence to regulatory and ethical standards to avoid a repeat of such scandals. Such procedures and processes should be built on a system of doing the right thing and have transparency, integrity, impartiality and respect at its core.

A 'virtuous or professional model' is predicated with a strong focus on decision making for the 'greater good' and the benefit for society. It is built on strong moral principles in upholding cultural and social norms and looking for ways and means of continuous improvement (Fewings 2009). An example is Environmental Ethics which is concerned with recusing carbon generation, tackling the climate emergency, promoting good health, reducing waste (especially plastics) and promoting more sustainable communities. Education clearly plays a major role in promoting such an approach and environmental and sustainability teaching and learning programmes in schools, colleges and universities are now commonplace in the developed world. In a construction industry context, there has been a recent push to consider the ways and means to reduce embodied in-building processes and utilise more environmentally sustainable materials and products. In addition, there has been a more concerted and integrated consideration of 'life cycle' costs rather than simply capital costs in the feasibility stages of projects. This is a welcomed move and considers the renewable technologies and energy-saving measures which could reduce the 'carbon footprint' of a building over its complete lifespan. The virtuous model also covers 'social and corporate responsibility' of organisations. This relates to doing the right thing for communities and the public at large, and having responsible policies on such aspects as employment, working conditions, health and safety, pollution and contributing towards local and national agenda.

An 'integrity and reputational model' is derived from the premise that organisations that create and build a strong ethical reputation over many years benefit from the public confidence that this brings for them and their brand. When organisations adopt this model, it is not always for altruistic reasons but linked to improving their image or creating positive public relations to increase their market share. An example in the UK could be Marks and Spencer, who has maintained for decades a reputation for providing good-quality products at a reasonable price. This integrity and reputational model can generate increased sales or work order for organisations and potentially increase profits in this way.

2.9 The Life Cycle of Buildings

The man-made environment which we live in, some referred to as the 'built environment'. In considering the built environment in which we develop and live within, one needs to consider this from the perspective of the life cycle of building and engineering structures. The life cycle in this regard does not purely look at the build period, known commonly as the construction contract period or construction phase, but extends to the operational/occupation period of buildings known as the 'in use phase' (Fewings 2009). Prior to the construction period, part of this notion of life cycle includes the development phases involving the inception, planning and tender stages of projects. Furthermore, the decommissioning and subsequent demolition of buildings and engineering structures can be regarded as the final phases of the life cycle, which may transcend into the life cycle of a new building.

The terms 'Whole Life Costs' (WLC) and 'Life Cycle Costs' (LCC) have been used interchangeably in the past and consequently their meanings have become confused.

Furthermore, the components of a whole life cost calculation have varied between clients, consultants and contractors. With no common ground, clients are not always sure what they are asking for, making comparisons impossible and the difficulty of working out whether actual costs match up to their target estimates. This unsatisfactory situation began to be addressed in June 2008, with the publication of two documents on life cycle costing: an international standard and a UK supplement. The international standard, BS/ ISO 15686-5 Buildings and Constructed Assets, sets out clear definitions for the two terms. In this document Whole Life Costing (WLC) is classified as a methodology for the systematic economic consideration of all whole life costs and benefits over a period of analysis. Another definition is given as 'an economic assessment considering all agreed projected significant and relevant cost flows over a period of analysis expressed in monetary value, and where the projected costs are those needed to achieve defined levels of performance, including reliability, safety and availability'. Life Cycle Cost (LCC) is described as the cost of an asset, or its part throughout its cycle life, while fulfilling the performance requirements. Furthermore, BS/ ISO 15686-5 Buildings and Constructed Assets: Service Life Planning: Life Cycle Costing Its UK supplement, provides a standardised method for life cycle costing for construction procurement. It clarifies the definitions for the UK market and sets down in detail how companies should go about working out a life cycle cost plan. Broadly, life cycle costs are those associated directly with constructing and operating the building; while whole life costs include other costs such as land, income from the building and support costs associated with the activity within the building. The expertise of the construction industry is best placed to deliver life cycle costs, which its clients can then use to calculate whole life costs. Agreement on these definitions and a consistent approach should enable life cycle costing and whole life costing to become more widespread. Figure 2.2 illustrates the components parts and relationship between WLC and LCC (adapted from BS ISO 15686-5).

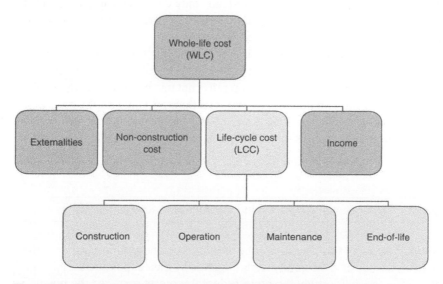

Figure 2.2 The components, parts and relationship between WLC and LCC.

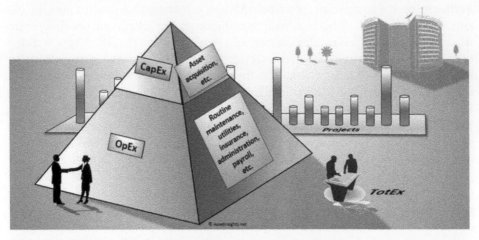

Figure 2.3   Operational and capital costs of a building. Assetinsights.net, 2021 / Asset Insights.

During the life cycle of buildings there are many different organisations and individuals involved with differing roles and responsibilities and these were articulated and discussed earlier in the chapter. There are also competing interests at large especially at the design and construction stages. Decision making at the early stages of the life cycle can have long-lasting effects on later phases such as the 'in use' occupation phase. It has been long accepted that the cost of a building's life cycle during the occupation, commonly referred to as operational costs or 'Opex Costs' normally represents approximately 80% of the total cost of a building over its life. Conversely, the costs during the development and construction phase, commonly referred to as capital costs, or 'Capex Costs' normally represent only 20% of the total. Operational and capital costs are illustrated in Figure 2.3. Despite this, it is normally only the initial capital costs of the build that are commonly considered in business cases as part of the decision-making process at the early financial viability stages.

Considering only the capital costs of buildings at feasibility stages can have negative connotations for projects whereby they could be compromised by too much emphasis on the initial construction costs. As one example, it is not uncommon for decisions to be made at these early stages to reduce the capital costs through 'value engineering' of the building design which could have long-term implications for the future performance, maintenance and running costs of the buildings. An example of this dilemma could be demonstrated below.

Example of early decision making affecting the long-term future of buildings

A client organisation has plans to redevelop their accommodation by procuring a new building. As part of this process, they develop a strong business case for the new build predicated on a feasibility study prepared by their appointed design consultants. In the feasibility study the projected total projects costs for the building were estimated as £20m, based on an elemental cost breakdown prepared by the appointed quantity surveyor acting as the cost consultant. During the next stages of design development when the full implications of the detailed design are being considered, and progressed, through consultation with end users it becomes apparent that the build is more complex and sophisticated than what was initially envisaged. The anticipated project costs at this later stage reflect this added complexity and are revised to £25m.

This increase in capital costs now creates a question for the client organisation around the financial viability of the increased value of the project since the anticipated business outcomes remain the same. A value engineering process is undertaken with the client and design team assessing where cost savings can be made on the build. A 'shopping list' of potentially desirable rather than essential scope of works is considered as part of this rationalisation process. Decisions are made to redesign certain elements of the build for more cost-effective solutions and in some cases certain items of work are omitted. The principal considerations at this stage are purely to 'balance the books' and make sufficient capital cost savings to make the scheme viable. There is little if any foresight or consideration of how these modifications and omissions will affect the performance, maintenance and running of the new facility.

Other considerations linked to sustainability are simply discounted on the basis of affordability. The decision making at this critical juncture over the life of the building has severe implications in restricting the performance of the facility and making unable to fulfill those initial anticipated outcomes articulated in the business plan. In addition, the cost of running the facility over the 30-year life of the building have increased owing to omission of renewable energy solutions increasing gas and electricity consumption.

Maintenance costs have also increased significantly as a result of less robust and arguably more inferior materials, components and building systems being installed. The cumulative effect of savings £5m on the capital costs and bringing the project 'back on budget' has in fact added an additional £40m over the 30-year life of the building in operational costs. This impact is further compounded by a view that the functional capability and capacity of the building has been compromised to the extent that it no longer performs to facilitate the operational improvements initially envisaged.

The above scenario hopefully demonstrates that early design making can have long-lasting consequences to not only the performance of buildings but can potentially impose overarching restrictions on organisational operations, with potentially severe implications for businesses. It also demonstrates that short-term planning can have long-lasting detrimental financial effects which could far outweigh the original cost saving. It is worth considering this scenario in the context of professional ethics as those decisions made by relatively few individuals at the design and construction phases can potentially affect many others in the occupational stages. There is also the notion that those making the decisions at the early stages to keep the capital costs within budget often compromise the build knowing full well that such decisions will be to the detriment of the build in the long term. Some would argue that these individuals could be compromised in the cost-cutting process and that this could be it represents unethical and immoral behaviour on their part. Others could suggest that pursuing strategies linked primarily with meeting capital affordability budgets at the expense of other aspects such as sustainability and specifically carbon generation, is short signed and misguided from an ethical standpoint.

In consideration of the above there is a long-standing view that the construction industry is predicated on meeting short-term needs rather than focusing on long-term sustainable solutions. Accordingly, it has been criticised for its general lack of longer-term future visioning for projects with insufficient emphasis at the decision-making stages on full life cycle considerations.

2.9.1 Life Cycle Analysis

Life cycle analysis is an assessment of the environmental impact of a product or service throughout its life cycle, from cradle-to-grave. The 'Green Guide' produced by the Building Research Establishment is a database of the Life cycle analysis of a variety of construction products. It rates each product on an A + to E ranking system, where A + represents the best environmental performance/least environmental impact, and E the worst environmental performance/most environmental impact. The environmental rankings are based on life cycle assessments using the Building Research Establishment's Environmental Profiles Methodology 2008. The Life Cycle Assessments of elements can be obtaining the Green Guide rating for the generic element or by obtaining validation of the environmental product declarations or other independently validated life cycle analysis available on the chosen products. Carrying out this whole life costing and life cycle analysis exercise enables project teams to demonstrate that they have considered the environmental and economic impacts of their decisions and chosen the best materials/products for the job. The option with the lowest discounted life cycle costing should normally be the preferred option assuming it results in lower energy consumption, reduction in maintenance, prolonged replacements or dismantling, recycling and re-use. Application of life cycle costing and whole cycle costing techniques provides clients and end-users with improved awareness of the factors that drive cost and the resources required by them for building. It is important that these cost drivers are identified so that most management effort is applied to the most cost-effective areas of the building. These techniques do not accurately predict the cost of occupying and operating the building over its life, especially at early stages, but they do allow economic judgements to be made between alternative technical solutions. Whole life costing should be carried out at different stages throughout projects from feasibility to detailed design stages. As the design progresses a more detailed life cycle / whole cycle costing analysis can be undertaken by incorporating actual manufacturers cost and life expectancy data. Where the choice of design solution provides for reductions in energy usage input from mechanical and electrical consultants will be required to calculate these reductions prior to a whole cycle costing calculation being carried out. Figure 2.4 illustrates life

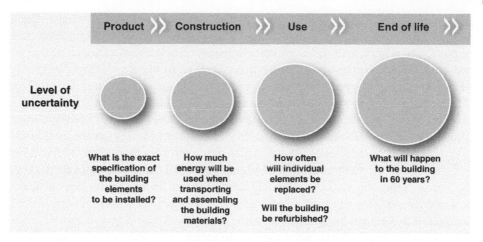

Figure 2.4   Life cycle analysis over the life of a building and varying levels of uncertain.

cycle analysis over the life of a building from cradle to grave and the varying degree of uncertainty that exists at different stages.

2.10   Summary

The construction industry accounts for a large percentage of the global economy and employs between 2% and 10% of the total workforce in most countries of the world. The performance of the industry can be seen to be a good indicator of the state and health of the wider economy. Despite the scale and importance of the UK construction industry, it is frequently criticised for performing poorly. This could be attributed to a general tendency for those employed in the construction industry to ignore best practice and be resistant to change.

The construction industry has emerged as a bespoke project-based industry where there are many different characteristics to other industries such as manufacturing. This is largely due to construction projects being nearly always unique and 'tailor made' to suit clients' individual requirements. This bespoke project-based industry is made up of mostly small teams, ranging from construction workers to design consultants. Such teams come together on a temporary basis for the life of a project and then disband to undertake different projects and this makes it very different from most other industry such as manufacturing. This also introduces fragmentation which can sometimes present an ever-increasing challenge for the sector. Owing to the fragmentation and bespoke nature of its composition, there are several unknown factors in any construction process which could cause the cost of projects to increase, programmes to be delayed and the quality of buildings to be compromised. This creates two aspects which come into play around commerciality and risk and who incurs any additional costs.

Project teams are made up of many construction professionals who have different roles and responsibilities. Clients are normally the project sponsors; accordingly they should be instrumental in steering, resourcing and leading all stages of the construction process from concept to completion. They appoint design teams which normally consist of project managers, architects, structural engineers, mechanical and electrical engineers and quantity surveyors. All these professional consultants work together closely within their different specialist areas during the design and construction phases. This coordinated and collaborative approach is designed to enhance teamwork for procuring successful project outcomes. On the contracting side of project team there are main contractors, subcontractors and suppliers. Main contractors are responsible for the day-to-day oversight of construction site activities which includes management of vendors and trades, and the communication of information to all involved parties throughout the course of building projects. Normally subcontractors are specialists in a particular area or trade, and they will perform all or part of the obligations of the main contractor's contract. It is common for a large construction project or renovation to involve many different subcontractors who work together to complete projects in a safe and timely manner.

The meaning of ethics can vary depending between the different roles and responsibilities of many different individuals that make up project teams. Clients for instance will expect ethical behaviours from their teams in managing projects closely and reporting any potential risks to them, as issues can frequently create additional cost and time delay implications for them. Conversely, from the consultant project teams perspective, there may be other ethical considerations to those of their clients. These may include expectations that clients are transparent in keeping them abreast of all issues especially around budgets and any issues that may affect progress of construction projects. From main contractors', subcontractors' and suppliers' perspectives, their view on ethics can be different again from clients and consultants. One of their main concerns is to be paid on time and to be treated fairly in commercial and contracting matters. Timely payments for these parties are especially important as cash flow for them is critical in their financial dealings.

There is an implied duty of care alongside established values and standards which govern whether decision making by construction professionals can be deemed good, bad or indifferent. However, there are sometimes problems for the construction industry in deciding how we judge what is right and wrong and therein what is acceptable. Models of investigation predicated on different branches of ethics can assist in deciding whether a particular situation or decision is ethical or unethical. These include the 'moral intensity', 'business', 'virtuous or professional' and 'integrity and reputational' models.

The life cycle of a building does not purely look at the build period, known commonly as the construction contract period or construction phase, but extends to the operational/occupation period also, known as the 'in use phase'. There are competing interests at large between the design and construction phase and the in-use phase, normally predicated on capital costs versus operational costs. Decision making at the early stages of the life cycle can have long-lasting effects on the occupation in-use phase. Operational costs during the in-use phase normally represents most of the total cost of a building over its life whereas the capital costs during the development and construction phase representing only a fraction of total costs. Despite this, projects can become compromised by too much emphasis on the initial construction costs where capital savings may be required to meet budgetary requirements. These savings as sometimes referred to as 'value engineering' and this short-term planning can have long-lasting detrimental financial effects, which could far outweigh the original cost saving.

Taking the different roles, responsibilities and perspectives that this chapter has discussed, the next chapter will articulate the significance and relevance of ethics for the construction industry with a focus on why ethics is so important to society.

References

Challender, J. (2017). Trust in collaborative construction procurement strategies. Proceedings of the Institution of Civil Engineers. Management Procurement and Law.

Challender, J. and Whittaker, R. (2019). *The Client Role in Successful Construction Projects*. Oxon: Routledge.

Egan, J. (1998). *Rethinking Construction. The Report of the Construction Task Force*. London: DETR.TSO. 18–20.

Egan, J. (2002). *Accelerating Change. Rethinking Construction*. London: Strategic Forum for Construction.

Farmer, M. (2016). *Modernise or Die: The Farmer Review of the UK Construction Labour Market*. London: Construction Leadership Council.

Fewings, P. (2009). *Ethics for the Built Environment*. London: Routledge.

Harrison, M.R. (2005). *An Introduction to Business and Management Ethics*. Basingstoke: Palgrave Macmillan.

HM Government (2013). *Construction 2025. Industry Strategy: Government and Industry in Partnership*. London: HM Government. 23–25. 61–71.

Inuwa, I.I., Usman, N.D., and Dantong, J.S.D. (2015). The effects of unethical professional practice on construction projects performance in Nigeria. *African Journal of Applied Research (AJAR)* 1 (1): 72–88.

Latham, M. (1994). *Constructing the Team*. London: The Stationery Office.

ONS (2020). Construction output in the UK. Office for National Statistics Construction. Available from http://www.ons.gov.uk [accessed Jan 2020].

Rahman, M.M., and Kumaraswamy, M.M. (2005). Rational selection for collaborative working arrangements. *Journal of Construction Engineering and Management* 131 (10): 1087–1098.

Rahman, H.A., Karim, S.B.A., Danuri, M.S.M., Berawi, M.A., and Wen, Y.X. (2007). Does professional ethics affect construction quality? *Quantity Surveying International Conference.4-5 September 2007, Kuala Lumpur, Malaysia*. P1–10

RICS (2012). NRM1, New Rules of Measurement. RICS.

Tabassi, A., Ramli, M., Hassan, A., and Bakar, A. (2011). Effects of training and motivation practices on teamwork improvement and improvement and task efficiency: The case of construction firms. *International Journal of Project Management* 30: 213–224.

3

The Significance and Relevance of Ethics

The character ethic, which I believe to be the foundation of success, teaches that there are basic principles of effective living, and that people can only experience true success and enduring happiness as they learn and integrate these principles into their basic character.

Stephen R. Covey

3.1 Introduction to the Chapter

This chapter of the book is specifically focused on the overriding question of why ethics is so important to society. As context it will attempt to define what ethics are but, in this pursuit, will articulate the ongoing dilemma of the different definitions of ethics. Such differentiation could pose a problem in the quest to understand and interpret what constitute ethics and ethical practices. Furthermore, as a foundation for the discussion on ethics, the origins and history of ethics will be explored from the philosophies and teachings of Socrates in the fifth century BC to modern-day doctrines on ethical frameworks. In this regard, historical theories will be explained and contextualised as the basis for ethical teaching over the centuries.

The chapter will also delve into the area of why ethics are important and will discuss the goals and benefits for individuals, organisations and society at large from ethical principles and values. In this regard, examples of unethical practices in the construction industry will be specifically discussed and the consequential effects and fall out from these in terms of reputational damage and loss of public trust. Finally, the important issue of how to recognise unethical practices and dilemmas about what is right and wrong will be explored and how to address such adversities. This is designed with the intention of changing cultures, addressing existing bad practices and improving working practices, especially in the construction industry, for the better.

3.2 What Are Professional Ethics?

To address the issues of professional ethics, particularly applied to the construction industry, one needs to firstly understand what ethics are and what constitutes ethical or non-ethical practices. Professionals are bound by a set of attitudes, principles and

Professional Ethics in Construction and Engineering, First Edition. Jason Challender.

character dispositions that govern the way their profession is practiced, and this is commonly referred to as 'professional ethics'. Such ethics are not confined to clients but according to obligations are also owed to colleagues and members of the public at large.

Theories of ethics come from a philosopher's perspective and can be categorized as metaethics, relating to where ethical values and principles emerge from, normative ethics, relating to moral standards of conduct and applied ethics, involving examining controversial issues (Internet Encyclopaedia of Philosophy 2010).

Ethics has been described in general usage as:

> ...*the philosophy of human conduct with an emphasis on moral questions of right and wrong' (Helgadottir 2008), 'the system of moral values by which the rights and wrongs of behaviour are judged' (Rosenthal and Rosnow 1991) and 'a moral philosophy that involves systematizing, defending and recommending concepts of right and wrong behaviour'.*

Alternatively, ethics could also be defined as:

> ...*the systematic attempt to make sense of individual, group, organizational, professional, social, market and global moral experience in such a way to determine the desirable, prioritized ends that are worth pursuing, the right rules and obligations that ought to govern human conduct, the virtuous intentions and character traits that deserve development in life and to act accordingly.*
>
> (Petrick and Quinn 2008 as cited in Helgadottir 2008)

A general definition of ethics was provided by Carey and Doherty (1968, as cited in Poon 2003) around the philosophy of human conduct with an emphasis on moral questions of right and wrong. Notwithstanding this, professional ethics has been described specifically around the expectations around responsibility, competence and willingness offer quality of services to the general public. There is sometimes confusion on what constitutes business and professional ethics. Accordingly, it would be helpful to examine the differentiation between the two. To offer clarity in this regard, business ethics are primarily focused around conduct. They revolve around ethical questions related to whether certain actions could be deemed good or bad, right or wrong, virtuous or vicious, worthy of praise or blame, reward or punishment. It seems that defining professional ethics is not an easy and straightforward task. To reinforce this position, Uff (2003) stated that 'in any event, it would be difficult today to pin down ethics to a particular definition in the current climate of change in matters of professional accountability and transparency'. Notwithstanding this challenge, professional ethics can be defined as 'a set of moral principles or values' (Trevino and Nelson 2004), whereas others have described them as a system of norms to deal with both the morality and behaviour of professionals in their daily practice (Abdul-Rahman et al. 2007). Alternatively, Fewings (2009) described professional ethics as the application of values to society and a range of different perspectives whereas Bayles (1989, cited in Abdul-Rahman et al. 2010) advocated that professional ethics revolve around a systems of norms wherein day-to-day behaviours and morality can be managed and regulated. Notwithstanding these alternative descriptions, when considering professional ethics, it is important to understand what 'profession' means. Profession in this sense could be defined as 'the possession and autonomous control of a body of specialist knowledge, which when combined with honorific status, confers power upon its holders' (Uff 2003). This is quite an arduous definition and as such it is probably more advantageous to consider professionalism from the perspective of providing a 'service' to clients. In this regard, a profession can be linked to a group of individuals who can provide specialist knowledge to their clients. Professional ethics therefore could be defined as a system of standards and norms to enable the behaviour and morality of those practicing professional to deal with their duties and job responsibilities. However, according to Hamzah (2012) professional ethics does not always apply solely to individuals but can be extended to organisations and institutions. Furthermore, one could argue that they are not solely for the benefit of clients but should be extended to the public. Moreover, they encompass qualities and duties associated with responsibility, competence abide by established standards, rules and behaviours.

When considering professional ethics, Lere (2003) described them as a way and approach between professionals and experts as experts and clients as lay people. Conversely when we examine professional ethics specifically within a construction context the RICS defined them as 'a set of moral principles extending beyond a formal code of conduct' (RICS 2010). In terms of how such moral principles should be applied in the workplace the code of ethics for project managers stresses that 'it is vital that Project Managers conduct their work in an ethical manner' (Walker 2009).

According to Rogers (1911) ethics revolves around the well-being of human beings. Well-being in this sense signifies the permanent realization of goodness enacted by individuals and takes into account:

- The nature of individual good
- The nature of social good
- The relationship between individual and social good
- The freedom of the will
- The ethical worth of positive morality
- The relationship between good and pleasure
- The nature of virtue (in antique ethics). In this sense virtue is linked with a person's character and applied to their motives or actions. In this context it is generally accepted that a morally virtuous person is one who abides and respects the moral codes laid down by honour, mercy, industry, temperance, charity. Conversely, the opposite of virtue is associated with vice.
- Duty and moral obligation (in modern ethics)
- The ethical motives that exist for people to pursue social good or to whatever is morally right

Another challenge for describing and defining ethics comes from the perspective that they have two distinct levels, namely Micro and Macro, relating to different aspects of the organisational environment. According to Chang (2005) micro ethics relates to personal issues and relationships with individuals and deals with issues such as personal integrity, honesty, trust and transparency. Alternatively, macro ethics deals with the wider aspects of actions affecting and impacting on society, the environment and the reputations of organisations (Chang 2005). Macro ethics can be related to ethical issues that have a significant and prolific high-level impact on the reputation of the industry or the public at large and predicated on the notion of doing good things for the wiser society.

Clearly the aforementioned alternative descriptions and definitions highlight that there are many differences of what ethics are which has given rise to problems of ambiguity and meaning. This view was supported by Vitell and Festervand (1987) who advocated that complications arise, and ethical dilemmas exist as there is no universally accepted definition of ethics. Perhaps this is one of the problems when considering professional ethics in the construction industry on a global scale which could be giving rise to different interpretations of ethics for construction professionals and what they mean in practice. This is arguably of particular importance when considering the construction industry as the understanding and views of many different professional bodies, association and organisations and their members are not always the same. The RICS Professional Ethics Working Party (RICS 2000) accepted that this potential ambiguity and inconsistency could create problems for the profession. Perhaps therefore a common framework for managing ethical dilemmas and thereby improving ethical standards is required which could address this problem. This opinion is supported by Liu et al. (2004) who argue that professional ethics is based on the subjective nature of principles, standards and values which vary between different sectors of the industry and a more consistent approach is required accordingly.

3.3 A Brief Historical and Theoretical Perspective on Ethics

The known history of pure ethics or moral (ethical) theories began with some of the ancient Greek philosophers such as Socrates, Plato and Aristotle. After recovery by English Positivists, ethics were the main area for discussions during medieval times in Europe. In more modern times throughout the nineteenth-century ethical ideas and theories were debated robustly throughout Europe. Evolution concepts emerged into physical sciences as well as the development of ethics during this time and were supported by the likes of Darwen, Comte and Spencer. Some of the main early philosophers and areas of ethics are summarised below:

(i) Socrates (fifth century BC)

Socrates was the founder of the Science of Ethics and professed the following moral beliefs and virtues:

- Learn your passions within your own soul and control them to reach wisdom
- He who knows must act accordingly
- No one voluntarily follows evil
- Virtue is knowledge
- Vice can only be because of ignorance
- Only by self-knowledge can freedom be acquired

Socrates believed that religious figures such as Zeus did not at all times act in a moral way. Accordingly, he advocated that people are not sure if behaviours are good because it pleases the gods or if it pleases the gods because it is good. Furthermore, he professed that some difficulties such as geometry can be solved through data whereas others including the justice system are moral issues.

(ii) Plato (fifth to fourth century BC)

Plato defined the social good and their relationships in his famous book of Republic. He advocated the four virtues belonging to the state which were justice, courage or fortitude and temperance. In this regard, justice was classed as the highest virtue and included the other three virtues. He defined the highest good in the form of ideas and reason in the universe and defined the notion of mortal body and immortal soul. He advocated that good people are ones whom knowledge, emotion and desire are in perfect harmony with the souls of themselves. He believed that people should give others the benefit of the doubt and not assume they have done wrong without sufficient evidence. Plato exclaimed that 'no one knowingly harms himself or does evil things to others because that would harm his soul'. Furthermore, Platonic ideals were predicated on justice being regarded as intuitive and arising from individual perceptions of forms.

(iii) Aristotle (fourth century BC)

Aristotle defined political science as the highest of all sciences, as everything else was aimed at the good of the state. Furthermore, 'social good' was considered by him above 'individual good' and this was related to individual actions attaining goodness for society. He defined well-being as the activity of the soul and it being in accordance with virtuous practices. Aristotle advocated that people should balance emotion and rationality and ethics should be predicated on moral choices and reasoned guiding actions with objectivism at its heart. Furthermore, his philosophies revolved around the notion that everything in the known universe has a beginning, end function, skill and purpose.

(iv) Early Christianity

Early Christianity focused on having a sense of one own personal goodness and morality and professed that this was the foundation of ethics. This stemmed from the underlying premise of 'love thy neighbour as thyself and do unto others as you would have them do unto you'. The early Christians also believed that quality is the basic principle of what they referred to as the 'golden rule'. Conversely, they proclaimed that anarchism related to people not wanting to be ruled as this implied to them inequality in society.

(v) Other historical perspectives on ethics

In addition to the above, Table 3.1 contains other historical perspectives and schools of thought around ethics.

Table 3.1 Other historical perspectives and schools of thought around ethics.

Ethical Perspective	Description
Hindu ethics	Hindu ethics were largely based on the qualities of virtue, truthfulness, self-restraint and inner purity. They were also predicated on the understanding that ethics cannot always be derived from first principles.
Daoist ethics	These were predicated on passivity leading to self-realization and that human nature is basically good.
Thomas Aquinas (13th century)	Thomas Aquinas practiced the notion that the reasoning of individuals naturally propels them to act in virtuous ways. However, he stressed that not all acts of virtuous acts necessary flow from natural inclination.
Thomas Hobbes (16th–17th century)	On the basis of social contract theory, Hobbes professed that not all things are objectively evil or good and people base good on what pleases them.
David Hume (18th century)	Hume argued that ethics are not always derived from rationality but more so on the way people feel. Furthermore, he professed that morality is inherent within the majority of the population. This, he argued, allows them to determine what is evil or good but conceded that rules are required due to the limitations of individuals.
Jeremy Bentham (18th–19th century)	Bentham studied and practiced Unitarianism and he preached that those actions which provide goodness for people could be considered to be righteous acts. He was a believer that 'the ends justify the means' in most cases and that the morality of a particular act could be judged not by its intentions but by the consequences. As such, ethical practices could be evaluated of the acts of people and specifically their behaviours which achieve goodness for others. He concurred that altruism in society should be as effective as possible to avoid or reduce the suffering of others and by the most effective way available to them.

3.4 Historical Theories as Frameworks for Ethics

In addition to the above historical perspectives the following theories have presented themselves in the past as theories as frameworks for ethical teaching.

(i) Deontology

Deontology was commonly perceived as a framework predicated on the notion that human beings have a propensity and duty to follow rules around ethics. Accordingly, there was a common understanding that people should treat others as they would be expected to be treated themselves. Deontology also revolved around the belief that good intentions and good will are still good even if their results do not achieve good results or consequences.

(ii) Social Contract Theory

Social Contract Theory was first introduced by Thomas Hobbes in the sixteenth century. The theory was grounded on the premise that people in their natural environment and habitat are constantly fearful of violence and death. In this regard, human lives were viewed as short, precarious, solitary and unpleasant. In an attempt to escape the miseries of these aforementioned predicaments people under Social Contract Theory doctrines were encouraged to relinquish powers they had over others in exchange for protection afforded to them from a strong authoritarian central medium. Protection in this regard was viewed as a social contract. Altruism and cooperation were regarded by Social Contract Theory as vital to create a system of ethics based on collectivistic and social ideologies.

(iii) Virtue ethics

Virtue ethics were predicated on the basis that people should behave in accordance with principles of virtue and be compassionate to others. Virtue is regarded in this sense as bringing pleasure and a peace of mind through doing the right thing and therein attaining the highest value.

(iv) Cognitivism

Cognitivism beliefs stemmed from the notion that propositions that are expressed from ethical statements can be either true or false. Conversely, non-cognitivism theories professed that ethical statements were the articulation and expression of opinions and emotions that were not always true to life. Accordingly, these expressions were considered to be more subjective than objective and neither true nor false in nature.

(v) Scientific ethics

Robert Merton in the 1950s advocated that scepticism existed around scientific theories and therein they were always open to challenge and questioning. In this regard, he philosophised that transparency and disclosure of results and data was not always practiced, with scientists trading intellectual property rights for esteem and recognition. He advocated a standard of Universalism for bringing about truth predicated on devising pre-established criteria. Following a similar belief, Bruno Latour in the late twentieth century was equally as sceptical of science, society and technology. He opined that 'scientific facts' were made up of social constructs designed to overwhelm the public by marshalling enough supporters and users. Over many years previous theories of the Universe have been proven to be wrong and those dismissed as wrong proven to be right. Latour also suggested that knowledge around science was sometimes floored as it is an artificial product of economic, political and social interactions in a competitive environment.

3.5 Concept and Purpose of Ethics

According to Guttmann (2006), ethics be designed to provide the mental powers to individuals to enable them to overcome fleeting passing instincts and therein enable them to choose good preferences over bad. Furthermore, Guttmann concurred that they should facilitate one's thought process in order that decisions emanate from the conscience of individuals to 'do the right thing'. Being ethical in this regard, he argued, should involve adopting moral codes, linked with values and principles. According to

Guttman doctrines, these are designed to ensure that rules and practices are consistently applied across a broad range of business situations. Such ethical practices should be conducted not only to clients but on society at large in all day-to-day business affairs. Guttman argued that it is important, especially for new construction professionals, to fully recognise what is required of them. This was predicated on the basis that professional conduct is not something that can be ignored and the way they conduct themselves is a major element of their recognition. Furthermore, Guttman professed that individuals should be taught to know right from wrong in their decision making and confident and assertive enough to uphold their actions and stand up to unethical behaviours.

According to Vee and Skitmore (2003) one of the important aspects of ethics is 'personal ethics'. This can be described as treating other people with the same extent of honesty and consideration as they would expect to be treated. Some professionals would concur that these personal ethics are more important for their clients than other individuals including members of the public.

3.6 A Context to the Discussion and Perception around Business Ethics

There is a perception that business ethics, best practice and codes of conduct are relatively things of the recent past, but some argue that the history of business ethics goes back to medieval times. According to the Roman Catholic Church, particularly in the medieval period, defined very carefully a piece of cannon law, which prescribed clearly what was legitimate behaviour in certain fields of the business world (Chryssides and Kaler 1993). There has been a marked increase in interest in business ethics in recent times; the subject has in fact quite a long pedigree. More recently and in the past few decades, firms and organisations have had more interests in business ethics than ever. Commitment to ethical business principles is becoming increasingly important (CMI 2010). There is little doubt that business ethics has become increasingly fashionable area of enquiry over recent years. (Mellahi and Wood 2003).

There seems to be a general agreement within the context of business that one of the fundamental values and distinctive characters of managing construction projects is acting on behalf and providing quality services to clients and stakeholders. Undertaking this role must be carried out professionally, ethically and within the rules and regulations of business engagement.

According to Interscape (2006) there are five different factors which influence ethics in the workplace:

- Individuals will tend to encompass more responsibility for managing their own careers at an earlier age
- Companies are adapting, innovating and changing which creates new ethical challenges
- It is becoming increasingly important in a global economy to understand other individuals' perspectives and positions on many different areas of work

- There is growing diversity throughout the world in matters such as religions, values and principles which raises the bar in a need for increased tolerance
- The complexity of knowledge and skills of the world's population creates new questions from an ethical perspective in terms of practice and technology

With reference to the above factors, these can create new challenges for the management of ethics where ethical dilemmas become commonplace. Particularly in the construction and engineering sectors the sensitivities to making the right choice and decision on ethical and morality grounds has become more prolific in recent years. Practices which may have been acceptable in previous years may now be regarded as inappropriate and at odds with professional behaviours. This could include accepting of gifts and hospitality but also the modern needs around transparency and accountability to many different stakeholders. Several research studies have been conducted throughout the years in the construction industry on the way they conduct their lives and working practices. Some quite alarming findings have emanated from these which have painted a less than good image of the sector. There have been reported cases of individuals prepared to lie and deceive to seal deals including new work orders, and in isolated cases people prepared to trick and deceive their colleagues and stakeholder to secure a competitive advantage. Other cases involved the use of company equipment and resources for their own gains and exaggerating and inflating time sheets and expenses for financial betterment.

3.7   Goals of Professional Ethics

According to Fewings (2009) goals of professional ethics can broadly be classified into two main categories; inward- and outward-facing goals, and these are listed below.

i) **Inward-facing goals**
   - Provides support and guidance for people to behave and act in an ethical manner, especially in instances where they are faced with dilemmas and pressures.
   - Creates common rules for people and organisations and outlines their responsibilities and duties to act and behave ethically.
   - Deters individuals to act unprofessionally and behave in an unethical way by creating an environment in which people are encouraged to report unethical behaviours and where sanctions are identified.
   - Reduces internal conflicts.
   - Creates standards of acceptable behaviours between colleagues, employees, employers, associates, clients and the general public.

ii) **Outward-facing goals**
   - Protects vulnerable populations who could be harmed by the activities of a profession.
   - Responds to previous cases of unethical conduct by professions.
   - Creates institutions that are resilient in the face of external pressures.
   - Serves as a platform for adjudicating disputes amongst professions and between members and non-members.

- Provides a platform for evaluation of professions and as a basis for public expectations.
- Establishes professions as distinct moral communities and worthy of autonomy from external control and regulation.
- Creates a basis for the development of trust and enhances and protects the good reputation of professions.

3.8 The Importance of Professional Ethics for the Construction Industry

In order to discuss professional ethics in context it is important to understand why it is important for the construction industry. Philosophers have debated for many centuries on the importance of ethics in society which has formed the basis of how the study of ethics awareness, education and standards has developed in the modern world. In this regard, perspectives have changed over the centuries, but ethical principles are largely built upon values, behaviours and trust in the society we live and work. Ethics has become a 'hot topic' in many sectors over recent years and the construction industry is by far no exception to this agenda. Notwithstanding this premise, reported cases of immoral and unethical practices and in some cases fraudulent criminal activities have given rise to some high-profile cases which have tainted the construction industry and put it in a bad light. This has had an overarching detrimental effect on the reputation of the industry and a distinct lack of public trust in construction professions and organisations. There have been reports of sleaze and corruption and ethical improprieties linked to collusion and regulatory breaches. Ethical teaching in applying professional standards and behaviours to the construction industry are vitally important to turning the tide and responding and addressing such unethical practices.

In the past there have been views that ethics are a vital and essential practice requirement as they engender the general public's trust and preserves their employers' interests (Abdul-Rahman et al. 2007). Construction professionals who act in an ethical manner will enhance their performance which will increase the success of projects. In addition, for construction professionals to survive it will demand public confidence which is dependent on the ethical conduct and professional knowledge (RICS 2010). Accordingly, this could suggest that any compromises on ethics could jeopardise the service delivery and damage public perception and the image of the construction industry.

One of the other problems the construction industry has faced over many years lies in its reputation and the general perception of the public (Robson 2000). This has largely emanated from negative press coverage, particularly in the UK, for unethical practices and behaviour uncovered in the past. From data published by the Building Research and Information Service Building it would seem that contractors in comparison with other construction professionals have gained a greater reputation for unethical behaviour. This assertion is predicated on a higher number of legal disputes emanating from their behaviour and practices over successive years (Vee and Skitmore 2003). Research studies conducted by Liu et al. (2004) suggested that developers and contractors especially display little emphasis on the existence of ethical codes when compared to other construction professionals such as architects and surveyors. This may therefore suggest that contractor-led organisation such as the Chartered Institute of Building (CIOB) should develop and promote cultural change to improve standards and raise awareness of ethical standards. Irrespective of whether the contractors are at fault, reputational damage to construction professionals may ensue when controversy on projects, culminating from their contractors' wrongdoing, emerges in the press. There have been many examples where clients have been held to account for the practices carried out by their appointed contractors. The example of Carillion's demise in 2018 in the UK led to speculation that public sector organisations, who awarded framework contracts for schools and hospitals, had been negligent in not undertaking more robust financial checks in their due diligence procedures.

A possibly more controversial explanation for unethical practices in contracting could be the relatively high turnover of construction company personnel and the perceived motivation of greed for increased profits (Liu et al. 2004). Construction companies are frequently accused of only being concerned with short-term economics or simply blinkered to the interests and well-being of their clients and associates in the industry. After all, why would construction companies become unduly concerned for their clients and practices when in most cases they move from one job to another in short succession? Arguably this is a misunderstood area in construction management. It can produce adversarial management styles geared to aggression and deception rather than a professional approach underpinned by integrity, honesty, transparency, fairness and trust (Walker 2009). Accordingly, this may explain why the importance and practice of professional ethics does not often feature very highly on the agenda of some building contractors. Construction professionals should be aware of such negative aspects and manage their selection processes to appoint contractors who have demonstrated social and moral responsibility in the past.

One previous example of controversy was related to construction workers being added to a database and blacklisted by up to 40 major national construction companies in 2003. It was reported that this could have excluded many individuals and companies from employment without their knowledge. This followed previous failures of standards and ethics related to The Office of Fair Trading's investigation and report of September 2007 which found that 103 construction firms had colluded with competitors in bid-rigging to secure construction contracts. This bid rigging practice is further discussed by Vee and Skitmore (2003) who explained that tendering has traditionally been the prime area of concern for the construction industry brought about by both contractor and awarding client unethical practices.

The aforementioned cases involving main contractors' misdemeanours does not infer by any means that other construction professionals including clients always comply with ethical codes. There have been cases in the past which involve construction clients, using their powers as 'paymasters' on projects, to pressure their professional consultants and main contractors to accept fixed price commissions and tenders which do not reflect reasonable margins for them to survive. Such win–lose scenarios, devoid of a partnering/teamwork philosophy could lead to reputational consequences for construction professionals under such scenarios. Clearly this is damaging to the construction industry at large in terms of public confidence and trust in the sector. While speculation exists as to the existing law on such matters these cases certainly highlight the debate on whether ethical standards and codes of conduct within the construction industry are being maintained.

The Chartered Management Institute articulate that ethical standards apply equally to the personal behaviour of individuals, as they do to organisations if their actions impact on society at large (CMI 2010). Furthermore, Trevino and Nelson (2004) described the importance of ethical standards from a personal perspective that staff need to feel that they are acting with integrity and in accordance with best practice. They concurred that individuals and professionals prefer to work for ethical organisations to feel good and be proud of the work they undertake. On a managerial level, organisations are responsible for the ethical or unethical practices of their employees under law. For this reason, it is important for employers to set down boundaries for acceptable and non-acceptable behaviours and practices of their staff (Trevino and Nelson 2004). This is especially relevant for managerial staff who should not abuse the authority and power of their positions and conduct their roles legally and legitimately. Walker (2009) argued that power can be a positive or negative force in its illegitimate form. Power in its positive form can be used to further objectives of the organisation, whereas in its negative form may be used to achieve personal objectives, which do not subscribe to organisational objectives.

One of the benefits of ethics for organisations revolves around trust, especially where firms are heavily reliant on their reputation for conducting business and gaining new work (Trevino and Nelson 2004). Weiss (2003) outlined that business ethics deals with three basic areas of managerial decision making. The first is around choices about the law and whether to follow, the second relates to choices about economics and social issues outside of law and the third is the priority of self over the company's interests. Despite these assertions, Cowton and Crisp (1998) argued that economies depend on the profit motive, and the pursuit of profit need not be perceived as immoral. To offer assistance to practitioners to what is moral or immoral Cowton and Crisp (1998) concurred that one needs to consider (i) respect for core human values, (ii) respect for local traditions and (iii) the belief that context matters to decide what is right and what is wrong.

Arguably, ethics can create the bedrock on which all of our relationships are built; it is about how we relate to our surroundings and it is not about the connection we have but the quality of that connection (Trevino and Nelson 2004). Another important aspect, which is at the heart of ethical and moral issues for construction professionals, is ethical dilemmas and decision making. It is recommended that there should be a need for sensitivity to ethical dilemmas to avoid actions being taken without the awareness of potential ethical issues. Chapter 4 of the book has been dedicated to such ethical dilemmas.

3.9 Ethical Principles and Codes for Construction Professionals

Owing to increasing concerns in many high-profile cases including those previously referred to in this chapter, demonstrating dishonesty and corruption, it is important for construction professionals to commit to and encourage project teams to comply with sustainable ethical principles. Codes of ethics which have been introduced have provided an indicator that organisations and institutions take ethical principles seriously as they should outline expectations for all personnel with regard to ethical behaviour and intolerance of unethical practices (CMI 2013).

Relationships between construction professionals and the professional consultants and contractors they appoint rely on professional ethics and trust especially since fee agreements cannot accurately specify all financial contract contingencies for possible additional services (Walker 2009). The main motivation for the public relying on members of professional bodies relates to them giving advice and practising in an ethical manner (RICS 2010). Accordingly, the RICS has developed 12 ethical principles to assist their members in maintaining professionalism and these relate to honesty, openness, transparency, accountability, objectivity, setting a good example, acting within one's own limitations and having the courage to make a stand. In order to maintain the integrity of the profession members are expected to have full commitment to these values. Furthermore, a 'Code of Ethics Checklist' published by the CMI sets down that ethics is particularly relevant to maintaining the reputation of an organisation and inspiring public confidence in it (CMI 2013). For this reason, codes of ethics should reflect the practices and cultures which construction professionals want to encourage for their respective organisations and project teams. This is supported by the CMI who advocated that:

> *A code of ethics is a statement of core values of an organisation and of the principles which guide the conduct and behaviour of the organisation and its employees in all their business activities.*

CMI (2013)

Arguably the main deficiencies of codes of ethics had emanated from the notion that there are no universal standards and accordingly they vary between countries and different sectors in the building industry. Boundaries and barriers created by fragmentation and differentiation within the construction sector have possibly deterred any common frameworks of professional ethics emerging in the past (Walker 2009). This is an area that demands more attention through multinational dialogue across all areas of the construction sector, to overcome. One attempt to address unethical behaviour in this way comes from The Global Infrastructure Act Anti-Corruption Centre, which has published a guide with examples of corruption in the infrastructure sector to assist practitioners. It sets down potentially criminal acts of fraud which include collusion, deception, bribes, cartels, extortion or similar offences at pre-qualification and tender, project execution and dispute resolution (Stansbury 2008). Furthermore Abdul-Rahman et al. (2007), from studies conducted on construction professional, published rankings for the top 11 most frequent unethical practices and these are contained in Table 3.2, and these include underbidding, bribery and collusion.

As a means to address some of these unethical practices, codes of conduct around professional ethics will be discussed further in Chapter 10.

Table 3.2 Top 11 most frequent unethical practices. Rank No. 1 = Most frequent. Rank No. 11 = Least frequent.

Ranking of Most Frequent Acts of Unethical Conduct	Ranking of Frequency
Under bidding, bid shopping, bid cutting	1
Bribery, corruption	2
Negligence	3
Front loading, claims game	4
Payment game	5
Unfair and dishonest conduct, fraud	6
Collusion	7
Conflict of interest	8
Change order game	9
Cover pricing, withdrawal of tender	10
Compensation of tendering cost	11

3.10 How Should Construction Professionals Recognise Unethical Practices?

The next issue and potential problem relates to what constitutes unethical behaviour in practice. As previously highlighted, there is no universal theory of ethics with different cultures existing within the construction industry and this creates problems and dilemmas in what is ethical or non-ethical (Liu et al. 2004). Clearly this reinforces the need for construction professionals to have a consistent approach to professional ethics which can be applied across the whole industry. Liu et al. (2004) explained, however, that this notion of achieving consistency is linked to the different cultures which exist within the construction industry and this may make boundaries between ethical and non-ethical behaviour become blurred at times. A practical example of this could include the boundary between receiving a seasonal gift as a polite gesture and what is deemed to constitute an act of bribery to influence an award of a contract for instance. Vee and Skitmore (2003) attempted to address this potential grey area and offered clarity on the boundary between gifts and bribery. They concluded that gift giving transfers become an illegal act of bribery when they compromise relationships between the gift giver and receiver and favour the interests of the gift giver. This is an important aspect for construction professionals, especially at tender stages when bidders may offer them gifts or invitations to corporate functions, to gain competitive advantage over their competitors. It is normal for construction professionals to have to sign up to anti-bribery legislation and declare gifts to avoid accusations of impropriety in such cases. Other forms of unethical behaviour could include breach of confidence, conflict of interest, fraudulent practices, deceit and trickery and may also in some cases have problems of grey areas and interpretation difficulties. Moreover, less obvious forms of unethical behaviour could include presenting unrealistic promises, exaggerating expertise, concealing design and construction errors or over-charging (Vee and Skitmore 2003). Such is the importance of this area; examples and types of unethical practices will be covered in more depth in Chapter 5.

3.11 The Need for Construction Professionals to Uphold Ethical and Cultural Values When Procuring Projects

> *Have the courage to say no. Have the courage to face the truth. Do the right thing because it is right. These are the magic keys to living your life with integrity.*
>
> W. Clement Stone

In consideration of the above quotation, the next important factor that arguably has a strong influence and link to professional ethics is culture and the cultural values of the construction industry and those construction client organisations that work within it. In the past, adversarial attitudes in the construction industry have affected relationships, behaviours, culture and trust in the construction industry. A major contributor in improving cultures with the construction industry has been professional ethics which defines rules of conduct. However, construction professionals should regard their scope and ethical responsibilities as much greater and more extensive than just simply concerning conduct, institutional rules and regulations (Walker 2009).

There have been opposing views on how best construction professionals can instigate cultural change within the industry and differentiation and fragmentation can again pose difficulties in this regard in aligning cultures, beliefs and standards (Liu et al. 2004). A practical example of this could be subcontractors having completely different standards beliefs and values to main contractors and similar scenarios existing between surveyors and architects. This raises the issue of the importance of changing the culture of the industry as a whole and the way it works. If cultural change is required, then a further question is raised as to how do construction professionals working with their project teams achieve this? The answer could be in improved training, education and personal development to raise the awareness of the importance of professional ethics. Ahrens (2004) certainly supported this view and advocated the use of modules designed to teach ethics to university-built environment undergraduates to expand their knowledge and understanding of ethical issues affecting the construction industry, with particular emphasis on contracting responsibilities and liabilities. He explained that too many young practitioners graduating from higher education do not possess the skills in areas relating to ethical values, moral working, cultural difference and environmental responsibility and is facilitating modules across 15 European Universities to attempt to address this educational imbalance. Liu et al. (2004) presented a similar argument to improve ethical self-regulation and cultural change through education and training. Further supporting views emanated from Vee and Skitmore (2003), who explained that ethical codes alone are not sufficient to maintain ethical conduct. Their findings indicated that employer-led training and institutional CPD to educate members on what ethical codes mean from a practical perspective can greatly increase awareness and participation in ethical practice.

3.12 Summary

This chapter has articulated the alternative descriptions of what ethics are which highlights that the differentiation and variation around definition could give rise to problems of ambiguity and meaning. Accordingly, an ethical dilemma could ensue from the absence of any universally accepted definition of ethics. Perhaps this is one of the problems when considering professional ethics in the construction industry on a global scale, possibly leading to varying interpretations of ethics for construction professionals and what they mean in practice.

Goals of professional ethics can broadly be classified into two main categories; inward- and outward-facing goals. Inward-facing goals include creating common rules for people and organisations and outlining their responsibilities and duties to act and behave ethically. Conversely, outward-facing goals include providing a platform for evaluation of professions and as a basis for public expectations.

The importance of ethics should not be underestimated as they are considered a vital and essential practice requirement, engendering the trust of the general public and preserving their employers' interests. Accordingly, one of the benefits of ethics for organisations revolves around trust, especially where firms are heavily reliant on their reputation for conducting business and gaining new work. Furthermore, construction professionals who act in an ethical manner will enhance their performance which will increase the success of projects. Conversely, any compromises on ethics could jeopardise service delivery and damage public perception and the image of the construction industry.

It is important for construction professionals to commit to and encourage project teams to comply with sustainable ethical principles but there are many different aspects and issues that influence and affect professional ethics, and this presents a challenge for the industry. Codes of ethics which have been introduced have provided an indicator that organisations and institutions take ethical principles seriously as they should outline expectations for all personnel with regard to ethical behaviour and intolerance of unethical practices. Notwithstanding this premise, unethical behaviours and practices have been experienced in the construction industry over many years. There are repercussions that can arise from non-ethical practices, especially in the context of the UK construction industry. Construction professionals, as leaders in the procurement of projects, should be leading the way in cultural changes to improve the reputation of the industry. In this pursuit, they should be aware of the importance of ethics, alternative definitions and interpretations of ethics, the reputation of the construction industry, codes of conduct and governance and regulations in avoiding bad practices. In conclusion, it would appear that measures to improve the practice of professional ethics, such as professional codes of conduct, have gone some way to improve the way the industry works but there are still far too many cases emerging of unethical practices that are blighting the sector. Although arguably these practices are emerging from a small minority of the sector, they are creating a bad press for the whole industry and further measures should be instigated by construction professionals to address this dilemma. Traditional responses in the past, at an institutional level, have been based on governance, regulations and punishment for non-compliance and clearly these have had only limited success. Perhaps construction professionals should be leading the way for a cultural change in the industry to train, educate and motivate construction individuals and organisations in what professional ethics entail, measures to ensure compliance and the benefits that they can bring for the sector. This could be achieved through more focus on FE and HE course modules linked to professional ethics and CPD through workshops and training events in the workplace. These measures will hopefully contribute to providing a more ethical environment for the industry to work within and reap great benefits not just for all construction-related organisations and the building projects that they procure, but for the future of the construction industry at large. It is accepted, however, that to bring about these cultural changes will take conviction, integrity and in some cases courage not to engage in established unethical practices. These improvements once ingrained within the industry could then reap massive rewards in providing a safer, honest, trusting and more enjoyable working environment for all.

References

Abdul-Rahman, H., Karim, S., Danuri, M., Berawi, M., and Wen, Y. (2007). Does professional ethics affect construction quality? Retrieved from www.sciencedirect.com [accessed 24th November 2019].

Abdul-Rahman, H., Wang, C., and Yap, X.W. (2010). How professional ethics impact construction quality: Perception and evidence in a fast-developing economy. *Scientific Research and Essays* 5 (23): 3742–3749.

Ahrens, C. (2004). Ethics in the built environment: A challenge for european universities. *ASEE Annual Conference Proceeding* (pp. 5281–5289).

Chang, C.M. (2005). Challenges for the new millennium. In: *Engineering Management* (ed. Pearson Education). Upper Saddle River, NJ: Pearson Educational.

Chryssides, G.D. and Kaler, J.H. (1993). An introduction to business ethics.

CMI (2010). Code of professional management practice. Retrieved October 12, 2010, from www.managers.org.uk/code/view-code-conduct [accessed 24th November 2019].

CMI (2013). *Codes of Ethics Checklist*. London: Chartered Management Institute.

Cowton, C. and Crisp, R. (1998). *Business Ethics: Perspective on the Practice of Theory*. Oxford: Oxford University Press.

Fewings, P. (2009). *Ethics for the Built Environment*. London: Routledge.

Guttmann, D. (2006). *Ethics in Social Work: A Context of Caring*. Philadelphia, PA: The Haworth Press.

Hamzah, A.R., Chen, W., and Wen Yap, X. (2012). How professional ethics impact construction quality: Perception and evidence in a fast-developing economy. *Scientific Research and Essays* 5 (23): 3742–3749.

Helgadottir, H. (2008). The ethical dimension of project management. *International Journal of Project Management* 26: 743–748.

Internet Encyclopaedia of Philosophy. (2010, March 24th). http://www.iep.utm.edu. Retrieved from Ethics and Self-Deception. [accessed 24th November 2019].

Interscape (2006). The World of Work is changing.

Lere, J.C. (2003). The impact of codes of ethics on decision making some insights from information economics. *Journal of Business Ethics* 48 (4): 365–379.

Liu, A.M.M., Fellows, R., and Nag, J. (2004). Surveyors' perspectives on ethics in organisational culture. *Engineering, Construction and Architectural Management* 11 (6): 438–449.

Mallahi, K. and Wood, G. (2003). *The Ethical Business: Challenges and Controversies*. Hampshire: Palgrave Macmillian.

Poon, J. (2003). Professional ethics for surveyors and construction project performance: What we need to know. *The RICS Foundation in Association with University of Wolverhampton* 148 (9): 232.

RICS (2000). *Guidance Notes on Professional Ethics*. London: RICS Professional Ethics Working Party.

RICS (2010). *Maintaining Professional and Ethical Standards*. London: RICS.

Robson, C. (2000). Ethics; a design responsibility. *Civil Engineering* 70 (1): 66–67.

Rogers, R.A.P. (1911). *Short History of Ethics*. London: Millan Books.

Rosenthal, R. and Rosnow, R.L. (1991). *Essentials of Behavioral Research Methods and Data Anlaysis*, 2nd ed. Boston: McGraw-Hill.

Stansbury, C.S. (2008). *Examples of Corruption in Infrastruture*. London: Global Infrastructure Anti-Corruption Centre.

Trevino, L.K. and Nelson, K.A. (2004). *Managing Business Ethics*, Third edition. USA: John Whiley and Sons.

Uff, J.P. (2003). Duties at the legal fringe: Ethics in construction law. *Society of Construction Law*, October 2010. Retrieved from www.scl.org.uk [accessed 24th November 2019].

Vee, C. and Skitmore, M. (2003). Professional ethics in the construction industry. *Engineering,Construction and Architectural Management* 10 (2): 117–127.

Vitell, C. and Festervand, D. (1987). Business ethics: Conflicts, practices and beliefs of industrial executives. *Journal of Business Ethics* 6: 111–122.

Walker, A. (2009). *Project Management in Construction*. Oxford: Blackwell Publishing Ltd.

Weiss, J.W. (2003). *Business Ethics: A Stakeholder and Issues Management Approach*, Third ed. Ohio: Thomson South-Western.

4

Ethical Dilemmas for Construction Practitioners

There is never a right way to do the wrong thing.
Parson (2005)

4.1 Introduction

The above quotation really typifies what ethics are all about. Notwithstanding measures to improve ethics in professional construction management, time and time again we have seen cases of unethical practices arise. These have taken the form of reverse actions and bid shopping, and according to Parson (2005), the construction industry is tainted by unethical behaviour. Accordingly, this chapter addresses the many dilemmas that can arise for construction professionals and offers practical guides and recommendations for maintaining ethical principles and practices under different scenarios.

In the last chapter the different roles, responsibilities and ethical considerations relating to construction professionals were discussed. Such differentiation in the nature of their roles and duties can from time to time create ethical dilemmas for these individuals, and these will be discussed. This chapter will also discuss how professional ethics are not always aligned to business ethics and the dilemmas that this can create, especially where commercial benefits are at stake. On this theme it will explain how such dilemmas can lead to non-adherence of codes of conduct and breaches of rules and regulatory standards. The importance and significance of ethical principles will be articulated alongside the benefits of upholding cultural values when procuring construction projects. In addition, factors influencing ethical dilemmas will be examined especially around the acceptability and appropriateness of accepting corporate gifts and hospitality. Regulation and governance policies and procedures within organisations are also discussed in this chapter which includes processes designed to avoid or address breaches including whistle blowing. The ethical dilemma of self-deception is discussed, and theories articulated of why individuals sometimes convince themselves that non-ethical practices are acceptable in some instances. It will also explain why the self-deceiver's warped perception of things may encourage and enable them to act in immoral and unethical ways.

One of the critical aspects of the construction process where ethical dilemmas can arise and adherence to fairness, morality and governance principles and processes is essential, occurs at the tendering stage of projects. Accordingly, this chapter presents

Professional Ethics in Construction and Engineering, First Edition. Jason Challender.
© 2022 John Wiley & Sons Ltd. Published 2022 by John Wiley & Sons Ltd.

different cases where potential breaches can arise at this stage leading up to the award of contracts. This is an attempt to steer readers on a path to maintaining robust ethical practices at all times, aligned to established codes of professional conduct. Finally, different scenarios and hypothetical examples are presented in this regard, throughout the chapter, together with the authors verdict of what is deemed to be the most appropriate action or decision in those cases.

4.2 The Construction Industry's Ethical Dilemma

Construction-related ethical dilemmas can be from a wide and different range of issues that may be to do with planning, conservation, waste management, land contamination, water irrigation, pollution, energy and many more issues. They can differ by nature, scale and effect but it seems they all have common dimensions within the context of professional ethics. Accordingly, it is vitally important that organisations and companies demonstrate their commitment to ethical and business principles (CMI 2010).

Despite the Catholics legislated for business ethics in the medieval times, other religious traditions have been equally concerned in their history with prescribing ways in which business affairs are to be conducted (Crryside and Kaler 1993). Reviewing some cases in recent histories perhaps begs the question of whether applying only codes of conduct, honesty and integrity enough? Furthermore, issuing warnings and advising only may not be enough to stop and prevent disasters. The space shuttle Challenger was tragically destroyed in 1986 during its launch, despite warnings by engineers, which were overridden by managers. According to Uff (2003) '...the debate continues as to whether the warnings were sufficient and whether the engineers should have prevented the fatal launch'. More recent cases including the Deepwater Horizon oil rig explosion in 2010 and the Grenfell Tower fire in 2017, where 72 lives were lost, are testament whether ethical lessons can be learnt to safeguard against further tragedies of this kind.

The issues around professional ethics affect a large spectrum of the population including private client organisations, local authorities, consultants, suppliers and contractors. According to Abdul-Rahman et al. (2010) construction professionals make judgements and use their own skills and knowledge as to their behaviours and the way they operate. They are ultimately responsible to their clients under their institution's professional code of ethics. Almost every profession globally has a code of ethics which is intended to provide a framework for establishing the correct and proper ethical choices and these will be discussed further in Chapter 7. Notwithstanding this premise, sometimes such professional ethics are not always aligned with business ethics, which are in the latter case predicated on making profits and meeting commercial targets of their respective organisations. In such cases, financial considerations can become a pertinent factor in an imperfect environment where competing factors can create ethical dilemmas. According to FMI this can be exasperated by fierce competition to win projects, low price mentalities and small contractor profit and overhead margins. Clearly this can create challenges and if not managed and regulated effectively, repercussions for the reputation of the construction industry.

According to Parson (2005) there is a large discrepancy between the value people place on ethics and what they actually do in practice to support their values. The common phraseology to describe this phenomenon would be that such people 'do not practice what they preach'. What reinforces this prognosis is that whilst a large majority of construction-related individuals believe that there need to be an industry-wide code of ethics, only a minority of them would agree that adding regulatory measures to enforce ethical behaviours is something they would want to introduce. Other substantiation for the same argument is that few construction companies still do not include ethical issues as part of their strategic plans or mission statements. Furthermore, some organisations have no ethical guidelines, policies or programmes for their staff to abide by. In such scenarios, it is perhaps not surprising that ethical considerations play 'second fiddle' to commercial matters and ethical breaches are commonplace in the construction industry dominated by profit margins.

Clearly this ethical dilemma needs to become 'centre stage' in the priority stakes for construction firms to take it seriously. Possible improvement measures and reforms could come from business leaders in setting thresholds within their respective organisations for personal ethical standards which could be reflected and applied down the line to their employees. Compliance with such ethical standards maybe something that could be enforced through human resource management (Parson 2005). This could take the form of training and education designed to make employees aware of what is acceptable and unacceptable practice and behaviours in the workplace. Implementation of measures and disciplinary procedures would need to be in place as a deterrent for breaches and to encourage good practice. However, this is easier said than done in practice. According to Parson (2005) levels of warnings for unethical behaviours would need to be managed together with disciplinary proceedings in certain cases. Processes to deter bad practices should also be in place. One example of this could be safeguards for financial and procurement processes to prevent impropriety and false accounting practices.

4.3 Ethical Principles and Values

It seems that ethical and moral principles are not as clear as black and white when considering whether certain behaviours are good or bad or a view on whether business dealings are right or wrong. For instance, some may refer to an increasing drive for increased profits as human nature, but others could perceive this as greed. The overriding consideration is perhaps not profiting per-say but the manner and morality they deploy in pursuit of such objectives. In an ever-increasing competitive world, there could be a misguided focus on self-interests, increasing market superiority and the battle for power. This has become more important during the Coronavirus pandemic where some construction-related organisations are struggling to survive and others seeking to take commercial advantage of a national crisis. The fallout from the Coronavirus pandemic and the ethical and unethical practices that have emerged will be discussed further in Chapter 8. In such instances, on both an individual and organisational level, this could result in cutting corners, disregarding ethical values and principles and failing to recognise the health, safety and well-being of subordinates, stakeholders and communities.

In light of above concerns, it is imperative that professionals working within the construction industry are to consider ethical principles and human values in their daily lives. Within the context of the law, they must work in compliance with legal and statutory requirements. Within the context of professionalism, they must adhere and follow their professional institutions' codes of conducts, which will be discussed further in Chapter 10. Within the context of humanity, they should and have to apply their principles, ethics and values as human beings. Examples to reflect the applications of above within the framework of the law are matters relating to health and safety, planning and environmental infringements. In the latter case, the job of environmental ethics is to outline our moral obligations in the face of such concerns (IEP 2010). On a purely professional level upholding the qualities, standards and codes of conduct of one's own institution, and taking decisions which can have long-lasting effects on stakeholders, communities and possibly humanity are all important ethical considerations. Other considerations include avoiding conflict of interests, safeguarding sensitive and confidential information, acting with integrity and honesty at all times, gaining trust and respects of others and maintaining equality, diversity and inclusion human resources policies. In addition, on a personal level, improving quality of own practice and competency through reflection and evaluation of own practice, continuing professional development (CPD) and completing personal development plans (PDP) are all prerequisites of good ethical practice.

Professional conduct and ethical duties of individuals should be applied in the wider public interests and not only confined to the interests of clients and employers (Uff 2003). Professionals, who are working in the construction industry, and often need to take actions and make decisions, must not only take into account the interests of their clients and stakeholders, but the wider communities and the environment as well. Fewings (2009) argued that personal values and ethics influence judgement on such actions and decision making. These in turn may have significant impacts and consequences on health, safety, social and life style issues and consequences can affect local, national or international communities.

4.4 Factors Affecting Ethical Dilemmas

According to Sutherland (1983) unethical conduct does not emanate from the upbringing of individuals but is more influenced by learning processes related to business practices. This could be significant as evidence has shown in the past that little continuous professional training (CPD) has been provided to employees in the general workplace. Taking account of this potential deficiency in teaching and learning could be the source of the dilemmas when looking at professional ethics. It could be giving rise to what some would describe as unacceptable levels of dishonest and unfair conduct in the construction industry. Conversely, Fewings (2009) opined that ethical dilemmas in the context of the busy world we live in can be created by competing demands. Examples of this could be where employees of young families are desperately trying to achieve a good 'work-life balance' but this is being constantly implicated by longer working hours brought about by increased workload and possibly the need to travel extensively as part of their careers. This can affect family life and could compromise the ability of individuals to perform well in the workplace. Such dilemmas in responsible organisations have resulted in policies linked to 'agile working' where individuals can work from home at certain time of the month. Other flexible working policies could involve employees being able to determine which hours they want to work whilst still honouring their weekly contract hours. For instance, some employees with childcare may choose to start early in order to finish earlier and collect their children from school. Others may simply prefer to shorten their working weeks by agreement of their employers to go part time.

There may be ethical dilemmas linked with employees conducting different roles where one role 'is at odds' with the other (Fewings 2009). An example could be a construction client who is trying to build collaborative partnership relationships with construction companies but being pressured by more senior executives to drive down the profits of those same companies to reduce costs on a project.

4.5 The Need for Construction Professionals to Uphold Cultural Values When Procuring Projects

Relativity applies to physics, not ethics.

Albert Einstein

In consideration of the above quotation, the next important factor that arguably has a strong influence and link to professional ethics is culture and the cultural values of the construction industry and those construction client organisations that work within it. Adversarial attitudes in the construction industry have affected relationships, behaviours, culture and trust in the construction industry (Challender et al. 2019). A major contributor in improving cultures with the construction industry has been professional ethics which defines rules of conduct. However, construction clients should regard their scope and ethical responsibilities as much greater and more extensive than just simply concerning conduct, institutional rules and regulations (Walker 2009).

There have been opposing views on how best construction clients can instigate cultural change within the industry and differentiation and fragmentation can again pose difficulties in this regard in aligning cultures, beliefs and standards (Liu et al. 2004). A practical example of this could be subcontractors having completely different standards beliefs and values to main contractors. Differentiation could exist between surveyors and architects in this way also given their different backgrounds, roles, responsibilities and priorities. This raises the issue of the importance of changing the culture of the industry as a whole and the way it works. If cultural change is required, then a further question is raised as to how do construction leaders working with their project teams achieve this? The answer could be in improved training, education and personal development to raise the awareness of the importance of professional ethics. Ahrens (2004) certainly supported this view and advocated the use of modules designed to teach ethics to university-built environment undergraduates to expand their knowledge and understanding of ethical issues affecting the construction industry, with particular emphasis on contracting responsibilities and liabilities. He explained that too many young practitioners graduating from higher education do not possess the skills in areas relating to ethical values, moral working, cultural difference and environmental responsibility and is facilitating modules across 15 European universities to attempt to address this educational imbalance. Liu et al. (2004) presented a similar argument to improve ethical self-regulation and cultural change through education and training. Further supporting views on this educational initiative emanated from Vee and Skitmore (2003), who explained that ethical codes alone are not sufficient to maintain ethical conduct. Their findings indicated that employer-led training and institutional continued professional development (CPD) to educate members on what ethical codes mean from a practical perspective can greatly increase awareness and participation in ethical practice.

4.6 Maintaining a Duty of Care in Tendering Processes

According to Uff (2003) questions frequently arise in the precontract stages and particularly during the tender processes for projects. With construction contracts being normally synonymous with large sums of money and profit for the successful bidder it is important to adopt best practices tendering procedures, compliant with the financial regulations of the awarding organisation and in the case of Europe, OJEU regulations. Any shortcomings in this regard could open the awarding body up for challenges from unsuccessful bidders especially where the awarding body is public sector. As part of the tender evaluation processes it is important to document each stage of the process, including the interviews and clarification sessions and treat all tender submissions equally. In the latter case, it is normal for pre-determined weighted criteria to be established in which each bid is scored against. This should be based not purely on cost as lowest cost selection rarely produces the best value (Latham 1994), but judged and adjudicated on quality related aspects such as experience, track record, specialist expertise and available resources. The tender evaluation process should seldom be undertaken by one individual but preferable by a panel of representatives. Sometimes these representatives are client managers and in other scenarios they consist of specialist consultants employed by clients. Arguably a mix of client and independent consultant involvement may be the best course of action, to add further robustness and rigour to the process, and defend any possible challenges. In some cases, anomalies may be uncovered during the tender evaluation process and this may relate to insufficient design or specification in the tender package. In such instances, questions may arise whether it is ethically justifiable to enter into a contract when one party is aware that major variations will be required or when there is insufficient information around such things as contamination and ground conditions. Other examples could be where the client's evaluation panel become aware that the successful bidding contractor has under-priced a particular project. This may be an arithmetical error or simple omission on the part of the contractor and there is a duty to make the contractor aware of these to stand by their tender or withdraw. The JCT (Joint Contracts Tribunal) code of practice for single-stage selective tendering has a contract procedure to deal with this eventuality. If this is not followed and the clients team seek to benefit from the tender anomalies it is highly likely in such cases that problems would come to fruition as the projects progress into the construction phases. This could result in financial claims, delays and possibly disputes arising and in extreme cases could lead to irrecoverable loss to the public as the beneficiary, and damage to the reputation of the construction industry. Accordingly, construction professions owe a duty of care and ethical responsibility to the public at large and not just their clients. In many cases this ethical responsibility is encapsulated within codes of conduct laid down by professional institutions, but in some cases existing codes do not provide sufficient guidance.

The tendering process can give rise to ethical dilemmas and issues which have in the past sometimes escalated into court proceedings, especially given that large sums of money are normally involved. Engaging construction law specialists does not exempt construction professional from following ethical practices and meeting minimum standards in this regard (Uff 2003). Accordingly, in any form of contract, construction professionals should not be able to hide behind their solicitors, especially when risks are being transferred. What can sometimes be difficult to gauge is identifying those standards and measuring against where those standards have been breached or complied with.

4.7 Corporate Hospitality and Gifts

There is no such thing as a free lunch.

John Ruskin

The above analogy can be applied within the context of professional ethics. One example relates to when is it acceptable to accept gift and/or corporate hospitality? There is no universal acceptance of what is acceptable or non-acceptable in all cases, but most organisations normally have rules of governance and declaration procedures to cover hospitality and particularly where gifts are involved. Such rules and procedures around governance should not, however, be seen as black and white. There may be instances where the self-deception of an individual eludes themselves into the belief that a gift or corporate hospitality will not affect their judgement or impartiality in a particular case (Parson 2005). An example could be a client representative who accepts a dinner invitation from a developer with the full knowledge and awareness that they will be interviewing them for a large construction project in the near future. There may be no organisational rules or regulations to preclude this, but one should consider how such actions could be perceived and whether it could give preferential treatment to that developer. This probably reinforced the importance of considering the circumstances and context around acceptance of corporate hospitality and gifts. One should consider the appropriateness of accepting a dinner invitation and what is deemed ethically acceptable or non-acceptable in such circumstances.

According to Fewings (2009) most organisations have clear guidelines on giving and, perhaps more commonly, receiving gifts and hospitality for the benefit of their employees. Other organisations have strict financial regulations which sets down requirements for declaring gifts and hospitality over a designated limit. Notwithstanding these limits, most larger organisations will prescribe that gifts and hospitality need to be declared on a central register which can be monitored and audited. Moreover, especially in the case of smaller companies with less resources around such matters as human resources, such matters are sometimes left to the employees themselves to decide whether they are acceptable and appropriate or not. This could be a dangerous and negligent position for companies to take and leave employees in a dilemma and susceptible to temptation in accepting gifts and hospitality offered to them (Fewings 2009). More responsible companies have formed guidelines and rules relating to what can be accepted and in what context. For instance, any gifts to a client who is empowered to award a building project from a contractor being considered would be totally inappropriate in all cases. Such an event would leave the gift taker vulnerable and susceptible to claims that their judgements were influenced by the gift bearer, and possible subsequent challenges from unsuccessful bidders. It can also lead to concealment of actions which can be toxic for any given organisation, and so transparency and declaration of gifts and hospitality coupled with clear boundaries for acceptance are the key to robust governance.

There are sometimes ethical dilemmas linked to the provision of gifts to existing or potential clients. An example could be a client representative receiving a 'small token of appreciation' from a contractor for their support in bringing them on a supplier framework. The next time around could involve a larger gift for a new work order. The question is could this gift giving be considered as a means to influence the client in giving the contractor future work orders? Some may consider that this could constitute a bribe in providing the motivation and incentivisation for future work opportunities being made available to that particular contractor. The other issue is whether this be deemed disadvantageous and unfair to the other contractors on the clients supply chain who may not receive the same privileges. This is where financial regulations should be unequivocal and clearly set down under the Bribery Act 2010 of what constitutes or more importantly what is not acceptable practice. For instance, some organisations lay down a maximum limit for receiving personal gifts and only in exceptional circumstance. One example could be a small Christmas gift as a thank you for support received throughout the year. In addition, most organisations have similar thresholds for receipt of hospitality. Once again, this is intended to provide a clear mandate on what can be received or not as the circumstances and value of the gift dictate. An invitation for a client to attend an award ceremony hosted by a contactor may be acceptable but depending on value may require to be declared under organisational finance regulations. If the same invite includes hotel accommodation, then this might 'tip the scale' in terms of acceptability.

From the examples already articulated, there is clearly an ethical dilemma that can arise from gift giving and receiving, and whether the gift in question to the scale of appropriateness and acceptability. To try to answer this question one needs to carefully assess the context and environment to which the gift is being offered within. In some countries of the world such as those in south east Asia, it is regarded as normal business practice, for gifts to be given and received between contractual parties as recognition of a new business relationship. Clearly the acceptability of the gift will depend very much on its value, but for modest gifts would be very much regarded as a 'token gesture' of goodwill. The opposite could be true of other countries in the world. In Denmark, a country recently ranked number one in the world for adopting the most ethical practices, any form of gifts of this nature could be regarded as highly inappropriate in business dealings. Most private and public sector organisations in Denmark have strict rules on not giving or accepting any gifts or hospitality whatsoever, as these could be seen as inducements or bribes.

Unfortunately, there is no universal guidance on what is acceptable or unacceptable in the area of when gifts and hospitality become bribery. Notwithstanding this premise, Fewings (2009), did attempt to provide a model for testing whether a gift or hospitality is appropriate and acceptable. The first question posed under this model related to whether the offering could be considered to be excessive. If the gift, by way of its value is deemed to be excessive then it should be rejected. This is especially the case when large contracts are being procured and the gift could be regarded as an inducement or influencing factor in the award. If the gift is not regarded as excessive and provided that its context cannot be linked to pending or future contracts, then normally it would be acceptable in that instance. The other consideration is whether the gift can be seen as a norm for the business being conducted. This addresses the cultural issues previously referred to where small gifts in countries such as China are an accepted part of everyday business. If the gift is regarded as not being a normal courtesy of business in a country, then once again it should not be accepted. The last question that should be addressed is related to whether the gift or hospitality could influence unethical conduct or behaviour. Where there is a pending contract to be awarded it would be regarded as highly inappropriate for any gift to be given by a potential benefactor to the contracting authority. Clearly the reason for this is to make a concerted effort to avoid associated said gift as being an influencing factor in the contract award. Any gifts that are received by a contracting authority at a time when they are looking to select and appoint a main contractor for instance, could be argued to influence their decision on the contractor and favour a particular contractor. Not only could this be regarded as unethical and unfair on the other competing contractors but could be regarded as a criminal activity under the banner of bribery. Bribery will be covered in Chapter 5 alongside other unethical conduct and corruption.

4.8 Regulation and Governance of Professional Ethics

According to Hamzah et al. (2010), one of the main dilemmas for the community of professional practitioners and professional institutions is related to self-regulation and autonomy. This can be called into question especially where clients and the public do not feel they are benefitting from the actions of professionals. In this regard they could feel disadvantaged and in extreme cases victims of unethical practices, and ethical quality control is called upon in such cases.

One of the issues for regulation emanates from the fragmentation of the construction industry in most countries. Construction extends across many different sectors including public and private sector clients including government agencies, developers, project managers, architects, engineers, surveyors, main contractors, sub-contractors, suppliers and manufacturers. When dealing with such a broad and diverse range of individuals and professions which play a part in the construction industry this creates the complexity of different skills, training and responsibilities. Furthermore, each profession is governed by their own professional institution which imposes their own bespoke professional codes of ethics, which will be discussed in Chapter 10. One could argue that some of the individuals that make up the construction industry fully practice professional ethics whilst the majority inclined to 'business ethics' (Hamzah et al. 2010). Main contractors and subcontractors who normally have commercial backgrounds and are faced with challenging profit margin targets could be more focused on business ethics than professional ethics. This dilemma is exacerbated by the fierce competition, low-price mentality and wafer-thin profit margins that are synonymous in the construction industry. This creates a dilemma for the industry where diversity creates an environment of competing interest and conflicting ethical practices and standards. Such a dilemma if unchecked could provide the platform for accountability, performance and quality to be compromised, to the detriment of clients and the public.

4.9 Whistle Blowing

Whistle blowing could be regarded as the procedure whereby employees can inform their respective organisations, normally anonymously or the public at large, of what they consider to be unethical or illegal behaviour of a colleague (Vee and Skitmore 2003). This can be a useful management practice for 'rooting out' inappropriate, unethical and potentially illegal behaviour in any organisation. The Public Information Disclosure Act 1998 provides employees protection in terms of whistle blowing and accordingly larger companies have policies specifically designed to allow staff to report unethical, immoral or illegal acts to the relevant representative. Such policies and procedure are designed to give staff the ability to report such practices and therein provide a more transparent workplace, without the fears of harassment or victimisation. In this way policies should encourage the practice of whistle blowing in order to address illegal or sub-standard practices. In the former case, where criminal activity could be a factor, the matter should also be reported to the police for addressing potential criminal action.

Whistle blowing from an ethical perspective should create a more open and transparent culture within organisations but is heavily reliant in a culture of confidential disclosure within organisations as opposed to a culture of fear, where acts go unreported and problems hidden. In the construction industry, many whistle blowing reporting has predominantly focused around breaches in construction health and safety and this will be covered more in Chapter 12. In cases where near misses have occurred, it is important for these to be reported in order to instigate improvement works and procedures to ensure that measures avoid the event reoccurring. Other common cases of whistle blowing have been related to theft, bullying, cover ups, corruption and bribery. It can be very damaging for organisations when such acts are exposed to the public. Consumer-related publications and news reports have uncovered extortion of some clients by contractors which has led to lengthy court cases and bad publicity for them.

In the UK, procedures around whistle blowing have been particularly applicable and utilised in the health sector, in a government drive to improve patient safety and care (Fewings 2009). However, it has sometimes been associated with breaches of confidentiality and in some extreme cases an abuse of the procedure, which can create ethical dilemmas in their own right. One example could be where one colleague maliciously claims that another has done something illegal under the protection of the claimant's anonymity. This could be a particular reprisal against an employee's line manager in cases where they feel that they have been unfairly treated by the individual. A hypothetical example of such a practice is detailed below:

Hypothetical example of an abuse of whistle-blowing protocols

A construction operative employed by a large national main contractor has been disciplined many times by a construction project manager who is his line manager. The disciplinary action has occurred over a prolonged period of time for bad time keeping, poor attendance and disruptive and aggressive behaviour on site. The individual is currently on a final warning and his line manager has warned him them any further violations would more than likely be classified as gross misconduct and most probably lead to dismissal. The relationship between the construction operative and the construction project manager has deteriorated as a product of the disciplinary proceedings. For this reason, the construction operative feels that he needs to do something to avoid dismissal and confuse the situation by making unfounded allegations about his line manager. Using his organisations whistle-blowing procedure, he raises an anonymous claim to the Human Relations department of his organisation that his line manager is bullying and intimidating various members of staff on site. He further outlines in the claim that the same individual is asking staff to conduct unsafe working practices on site, which is risking lives. The construction operative knows that all the allegations have been fabricated and therein false but is motivated to enact revenge on his line manager for disciplining him.

Under the whistle-blowing procedures the claims are treated seriously and owing to claims around serious health and safety violations, the construction project manager is suspended pending an internal investigation. An internal investigation is indeed conducted involving interviews with key site individuals, involving considerable time and cost. The investigation confirms that there is no case to answer, as no evidence to suggest any infringements have occurred. The construction project manager is reinstated after two months suspension but feels aggrieved that there has been a malicious and unfounded action against him and one that has caused him anxiety and reputational damage. Despite the whistle-blowing matter being closed, the construction project manager, suspecting that the claims were made by the construction operative, feels that he can no longer pursue disciplinary action against him for fear of another whistle-blowing claim. There is no incrimination on

---

(Continued)

(Continued)

the construction operative for raising the claim under the whistle-blowing procedure, and his identity remains anonymous. Furthermore, the construction operative's unethical actions have secured his position within the organisation owing to his line managers fears on pursuing any further disciplinary action against him.

Author's verdict

This clearly represents a case where a procedure introduced to prevent unethical actions and behaviours has been abused to reward unethical behaviour by an individual. The anonymous nature of the whistle-blowing procedure was designed to encourage staff to report unethical or illegal practices but, in this instance, has served to allow an individual to make damaging and false claims without any come back or accountability. This represents an issue for such whistle-blowing policies and developing protocols to safeguard against abuses of this kind.

4.10 Ethical Dilemmas around Self-Deception

Self-deception gives rise to many questions linked to morality and ethics. According to IEP (2010) it can stem from the self-deceivers distorted view of the world and of themselves. There are some who believe that the self-deceiver's warped perception of things may encourage and enable them to act in immoral and unethical ways. The ethics of belief presented a moral duty to form beliefs in response to all available evidence, and not primarily on what is convenient, comfortable and desirable. An example on this aspect could be where a building owner brings people onto their premises, in the full knowledge that there are health and safety concerns, to avoid any disruption to the running of their organisation. In such an example the individual may subconsciously stifle their doubts about the safety of the building to bring them to a conclusion that the building is safe. As a result of bringing people which may include contractors into an unsafe building, this could bring about injury or fatalities, whereupon the building owner would be held liable. Even if no injuries occurred, the building owner's moral status would still be the same, albeit they may not have been found out under this scenario.

There are also views about the corrosive effect of self-deception upon the belief and ability to 'test for truth'. Self-deception could affect an individual's 'belief-forming' processes. The easiest way a person might do this is by entering into a situation with their proverbial 'eyes closed'. An example could be when someone has an addiction, such as gambling, drinking or smoking, when they delude themselves into thinking that there are no issues and conducting themselves in a normal way. They may fool themselves in thinking that such pastimes are practiced by a large percentage of the population and therefore a perfectly normal activity. What they may fail to consider is the extent of extreme practices which may render these at odds with normal practice. For instance, there is a big difference to the consumption of an occasional beer amongst friends to someone who is alcohol dependant and drinking every hour of the day. Self-deception in a similar way could involve conducting unethical actions and behaviour to gain a financial or other advantage but deluding themselves that they are doing anything wrong in the process. This may help to explain why there are so many different cases of unethical behaviour in the construction industry where competition remains high, large sums of money are at stake and potential advantages sought for conducting unethical actions.

According to Hamzah et al. (2010) there is a real risk of people being less than sensitive or even blasé to ethical dilemmas, possibly linked to self-deception in the course of their lives and business dealings. In certain cases, this could lead them into scenarios where they are unaware that they have acted unethically through their actions or decision making. This is likely to result in them considering their behaviour as 'only marginally unethical' and in their own minds acceptable. An example could be in them telling a 'white lie' to one of their colleagues or clients to justify an action or outcome. The next time around could involve a more serious and complex derogation of the truth to cover their tracks possibly related to an act of negligence. The question is when does a 'bending of the truth' become an unethical act? Most professional institutions would say it is not acceptable for their members to lie in any case within a business context.

4.11 Practical Examples of Ethical Dilemmas

There are clear objective practical examples where behaviours, practices and decision making are either ethical or non-ethical. In a similar way it may be straightforward to deem where an action or behaviour is right or wrong where a 'black or white' scenario exists. However, there are arguably more frequent cases where there are many different 'shades of grey' around practices and decision-making scenarios which can be open to interpretation about what is right or wrong. This subjectivity creates a dilemma for the teaching and practice of ethics in knowing what is acceptable or unacceptable in a lot of cases. The context to the particular situation and the roles and responsibilities of the individuals need to be carefully considered alongside their seniority and decision-making authorities. One could argue that the more senior the individual and the increased authority for decision making which this brings, the more important ethical adherence becomes. For instance, it is likely that a senior construction responsible for awarding major capital projects to design teams and main contractors will have a higher level of ethical scrutiny and responsibility than a self-employed sub-contractor working on the same project. In this regard, the self-employed individual will have limited scope for decision making and possibly no responsibility for managing others. By contrast, the client will have responsibilities around procurement of the consultant and main contracting organisations involve in the build, which should be in adherence to ethical standards and regulatory requirement. This is more prominent in the public sector where governance and regulations especially around tendering and procurement are abundant. In the UK, public sector projects over £4.1m are subject to procedural rules set down by the EU and referred to as OJEU requirements. These stipulate how tendering needs to be administered and are designed to instil fairness and transparency into the award of major contracts. Fewings (2009) argued that it is ensuring that the decision-making processes are conducted in an ethic way that it is important rather than the outcome of the decision. A hypothetical example of this can be demonstrated below:

Hypothetical example of a project that led to poor outcomes but was managed in a professional and competent way

This example related to the award of a major construction contract by a City Council in the UK. The process was conducted in strict accordance with EU procurement regulations and the awarding Council's strict corporate governance requirements as laid down by their policies and Finance and Resources Committee. This involved a series of detailed procurement stages linked to pre-qualification and evaluation bids through pre-determined weighted criteria. The decision-making process on the award was taken by an assessment and evaluation panel purely on the submissions received and not related in any way to their own experiences of the successful or unsuccessful bidders. An agreement mechanism for the award of contract was agreed prior to tender action and a whole series of 'checks and balances' for avoidance of any financial irregularities. Despite this the main contractor that was awarded the contract fails to perform well with consequential delays to project, overspend and quality control issues. One would not dispute that the City Council has behaved in a professional and ethical way in awarding the contract on the basis of the submissions they received. They may have subconsciously preferred to award to another contractor who they have has prior successful dealings with but have consciously isolated their own preferences in this regard to ensure they are completely unbiased in the process of awarding a multimillion project.

Author's verdict

The strict adherence to the EU procurement rules coupled with the organisation's financial regulations has demonstrated that the awarding authority, in this case the City Council, has fulfilled its role and responsibilities in an ethical manner. Should there have been any challenges to the award of contract from one of the unsuccessful bidders then the process would inevitably have come under scrutiny. In this particular case, as a consequence of processes being strictly adhered to, then any challenge would likely to have been unsuccessful and the award deemed fair and equitable. The City Council will have had to produce documentation as evidence to support a robust and transparent procurement process. The fact that the contractor that has been awarded the project had failed to perform is incidental to the appointment process. This could not have been reasonably predicted at the award stage especially since the client body has proved they have been due diligence in the tender evaluation. The tender process had involved them in closely scrutinising the bids and investigating their suitability to the nature of the particular project, against predetermined assessment criteria. Investigation into the successful contractor's previous track record and achievements, alongside their approach to the project, the competence of their proposed project team and the resources they committed to deploy on the project had been careful considered by the assessment panel. Furthermore, the City Council had sought references and undertaken financial checks on the main contractor which had confirmed that they had a successful track record and financially sound.

Although the contractor did not achieve the successful outcomes that were intended, the process in the author's opinion was deemed to have been managed correctly. This example demonstrates that ethical processes are equally as important or in some case more important than the outcomes. Conversely, no one would condone successful outcomes brought about by unethical behaviours and short-circuiting of due process and procedures. This reinforces the perspective that delivering good outcomes does not justify unacceptable behaviour.

Sometimes ethical dilemmas can arise from individuals seeing their roles and responsibilities differently. One example of this could be that senior staff feel in certain instances that they can circumnavigate certain rules or regulations within the workplace. This can give rise to the old adage 'that rule does not apply to me because of my position in the organisation' or 'rules are made to be broken'. Other examples could be that some individuals purposefully interpret rules and regulations differently than they were intended. One example could relate to financial regulations requiring tendering procedures on all projects and senior staff choosing to justify not doing so in every case based on justifications such as time constraints. This create an ethical dilemma from the perspective of differentiation. If there are rules and regulations should they not apply to all staff rather than just a few? In addition, if some staff, especially senior staff with positions of influence and power, fragrantly breach processes and procedures then it can be detrimental for others to have the motivation to follow them. A vicious circle can result in some cases resulting in an organisational culture geared around avoidance and non-abiding by rules on the premise they are more guidelines than mandatory requirements. In such cases measures designed to safeguard against breaches are bypassed.

Other ethical dilemmas in practice could relate to a conflict of interest between business, managerial and professional complexities. There could be, for instance, a situation where a business decision in the interests of an employee conflicts with upholding the values of their profession. Consider the following hypothetical example:

Hypothetical example of an ethical dilemma related to conflict between client instructions and professional ethics

A quantity surveyor working for a client developer notices that a final account submitted by a main contractor has failed to include elements of work that they know to have been instructed as part of the project. In addition, the contractor has made arithmetic errors in not adding up their elements of work correctly. The omission of work elements has presumably been a mistake on the part of the contractor in failing to include those completed work packages. In addition, there is clearly an arithmetical error which has further disadvantaged the contractor with the overall final

(Continued)

(Continued)

account total being lower than it should have been. The quantity surveyor reports this anomaly to their clients in their financial forecast alongside recommendation to make the contractor aware of their mistakes. Subsequent to being made of this issue, the client instructs their appointed quantity surveyor to withhold the omissions and numerical inaccuracies on the basis that they will benefit from a lower final account if they remain silent. The client justifies this on the premise that 'it is their mistake and for them and only them to identify and correct'.

What is the correct course of action in this instance? Author's verdict

The quantity surveyor has become compromised in whether they should abide by the express instructions of their client or abide by their professional institution's codes of conduct for ethical practice. Notwithstanding their client's instructions, it is clear and apparent to the quantity surveyor that concealing mistakes, which they know will financially disadvantage the contractor, could be regarded as unethical practice on their part. The quantity surveyor also should consider that as a chartered surveyor they are bound by the ethical codes of practice laid down by the Royal Institution of Chartered Surveyors and rules and regulations therein. In consideration of those professional duties they should inform their clients that they have no alternative but to disclose the mistake to the contractor and allow the contractor to resubmit their final account for a higher value. This could bring the quantity surveyor into a potential different type of conflict with their clients, for not complying with their instructions. This case articulates that professional ethics should never be compromised by others instructing them to compromise for financial or other gains. An individual's professional ethics should not be influenced by stakeholders or external forces.

4.12   Ethical Dilemmas that Apply to Different Roles

The differentiation of roles and positions within the construction industry can strongly influence the type of ethical dilemmas each professional will be required to address. A few examples below could articulate how different members of project teams face different challenges and ethical dilemmas.

Main contractor and subcontractor ethical dilemmas

Both main contractors and subcontractors may have dilemmas relating to the quality of building that they are constructing balanced against the profit they are motivated to make on projects. For instance, if they wish to increase their profit margins on a construction project, they may decide to cut corners and use inferior materials and products. Arguably, such unethical practices are more prevalent on design and build contractors where contractors and their supply chain have responsibilities for the detail design and specifications of buildings. In these instances, the contractors will formulate their tender priced based, normally as part of a two-stage process, and seek to win contracts on a 'best value' basis, which sometimes can mean lowest cost. If they are so inclined thereafter, contractors can employ the ways and means to reduce their costs through 'value engineering' processes. This sometimes means omitting more expensive components of the build or simply substituting them for more inferior and less expensive elements. As the mechanical and electrical systems on building account for a large percentage of the overall costs and quality of installations vary considerably, this is a common area for contractors to recoup cost savings. This can lead to quality control issue for clients over the life cycle of buildings, sometimes leading to more onerous maintenance responsibilities and increased running/operational costs. Sometimes clients are oblivious at the handover stage of projects that such an approach has been instrumental in the design and construction phase albeit they become acutely aware of issues later on when latent defects arise, and parts of the build are compromised or fail. Arguably with traditionally procured projects it is less likely that they will become able to compromise in this way as clients and their design team will have more control over the specification and design process and will closely monitor that standards and adherence to specification are being strictly adhered to.

Another dilemma for both main contractors and subcontractors, relates to them attempting to increase profit levels or offset losses on a project by generating unfair and unreasonably contract claims for variations or disruption of the works. In certain instance this could involve them overcharging clients for additional works brought about by clients as the project develops. This is sometimes referred to as 'scope creep' and for this reason, from the clients' perspective should be avoided or at least minimised to a very low level of changes. However, sometimes contractor claims are not generated from such client changes. More unscrupulous contractors, albeit a minority, have been known to generate claims based on assumptions built into their tender prices which have not been realised. An example could be claims relating to savings contractors were hoping to make on disposal of contaminated ground following excavation works which have not transpired. It may be that additional costs are claimed even though ground disposal costs have in fact not increased. Other claims could be generated from inaccurate or exaggerated data relating to such uncontrollable events such as inclement weather, where clients are responsible for paying contactors extension of time with possible prolongation costs.

Other unethical dilemmas for main contractor involved what is commonly referred to as 'bid auctioning'. This involves them approaching one of their frequently used subcontractors and requesting that they reduce their prices to enable the main contractor to recoup any loses or increase their profit on the same building project. Whilst this is not illegal there are those that would suggest this is an unethical act as it disadvantages the subcontractor and places the burden on the subcontractor for resolving problems not of their making.

Ethical dilemmas for designers

Design teams on construction projects are made up of many different disciplines including architects, mechanical and electrical engineers, structural engineers and specialist designers e.g. acoustics and environmental engineers. These designers are constantly challenged by their clients to be appointed on a fixed fee basis with sometimes very little information to base their initial fee quotation upon. Accordingly, these professionals are constrained by the amount of time they can expend on a particular project based on their competitive fee in securing the appointment. For this reason, it is not uncommon that design consultants can under-estimate the amount of resources that a particular project will entail, especially when the scope of their services is unclear at the start.

This under-estimate may have resulted from the project being more complicated than they had envisaged and therein requiring more time and resources than they had originally allowed for within their fee proposals. Under such scenarios, designers have three main options:

Option 1

The first option relates to the designers investing the full time and resources that the project requires and honouring their original fee proposals. This represents the most ethical option but may mean they make no profit on the commission or even worse make a loss, when considering the amount of time expended against the fee received.

Option 2

The second option relates to expending the full time and resources as the first option but recharging the client for the additional fees incurred. Clients may dispute this additional cost if the consultant designers have agreed a fixed fee contract, rather than a 'time charge' agreement. Under this option it may be regarded as unethical to expect clients to pay additional fees for a project that the consultants have clearly under-priced. The exception to this would be when the consultant has clearly set down in their fee quotation the scope of services they have included for and this is extended within the contract through 'scope creep'. However, in these circumstances the consultant should make the client aware in writing as soon as additional fees are identified, and have clients agree to these before commencing additional work. What can be sometimes frustrating for clients is to be presented with increased fees for consultant services at the end of a project, which they have not been made aware of and therein are unexpected.

Option 3

The third option is probably regarded as the most unethical option and this relates to the consultant designers simply expending the same time and resources on a project as set down in their fee agreement, when they know that more time and resources are actually required on the project. This could involve taking short cuts in the design process or simply not designing certain elements and leaving certain elements of the design for the contractors to address. This can result in poor quality standards in some cases and scenarios that lead to defects and workmanship problems. Such problems can increase the 'life cycle costs' of buildings for clients and should be avoided at all costs. This scenario could be especially applicable to mechanical and electrical designs where installations do not meet industry standards and faults and poor environmental controls could be created. This option is both unethical practice and unfair on clients as they may not be aware that any resourcing issues are prevalent and as such oblivious that scaling back of design resources on their projects is taking place, causing them untold difficulties in the future.

Client ethical dilemmas

There has been much academic research conducted in recent years around clients awarding building contracts and consultant appointments on a lowest price basis rather than through best value evaluations. The main difference is that the lowest cost option does not take into account the quality of the contractor or consultant submissions and the clients assessment on whether these contractors and consultants can deliver projects successfully. Best value evaluations usually take into account the contractor or consultants track record, reputation, financial stability, expertise, experience and the team they have put forward in their tender submissions. Taking these aspects into account should in theory result in the most appropriate bids being accepted and reduce the risks that projects will fail. Simply accepting the lowest priced tender is synonymous with poor project outcomes (Latham 1994). Failure in these cases could mean delays to programmes, additional costs or buildings being compromised by poor quality and performance standards and poor site management (possibly leading to health and safety on site issues).

Notwithstanding the above premise, what can sometimes present a dilemma for clients is when they cannot resist the temptation to accept the lowest bid. Sometimes arises when they know that it is far lower than other bids received, and the bid does not give any confidence that the contractor or consultant can deliver the project successfully. In these circumstances, clients fool themselves into believing they can pay an unusually low price for a project and receive the same successful outcomes as other more robust and realistic bids.

Taking the above into account, there is an ethical question of whether clients should accept a relatively inferior bid based on price alone and expect their contractor or consultants to meet all the aims and objectives for the project. Normally when bids are submitted at an unusually low price, this could mean that the bidders have under-estimated the amount of work required or simply 'bought the job' with the intention of claiming for additional monies as the contract progresses. With profit margins for contractors and consultants, certainly in the UK being relatively low, this

(Continued)

(Continued)

could inevitably result in disputes between clients and their contracting organisations around claims for variations. It is unlikely that contractors and consultants will be able to sustain loses on such contracts and accordingly they will seek redress from clients to recoup monies that should have been included in their tender submissions. For clients, they should adopt ethical principles at the tender stages of contracts and of bids seem on the surface to 'be too good to be true' they normally are. This is the main cause of contract disputes where clients award contracts on purely financial grounds with little or no consideration on the bidder's capability to deliver on time, budget and performance.

The other ethical dilemmas for clients could relate to clients instructing addition works and refusing to pay contractors/consultants for the increased costs. As 'paymasters' on the contract clients sometime been known to coerce their contracting organisations into accepting terms and conditions which they feel forced to accept on the basis of future work order. This again could be regarded as 'bullying' behaviours and therein categorised as unethical practices and behaviours.

4.13 Summary

Construction-related ethical dilemmas can emanate from a variety of different issues that revolve around financial considerations, planning, conservation, waste management, land contamination, water, irrigation, pollution and energy. Professional ethics are not always aligned with business ethics, and there are sometimes conflicts with doing the right thing and making profits and achieving project targets. In such cases, financial considerations can become a pertinent factor in an imperfect environment where competing factors can create ethical challenges.

Few construction companies still do not include ethical issues as part of their strategic plans or mission statements. Furthermore, some organisations still have no ethical guidelines, policies or programmes for their staff to abide by. Possible improvement measures and reforms could come from business leaders in setting thresholds within their respective organisations for personal ethical standards which could be reflected and applied down the line to their employees. Compliance with such ethical standards maybe something that could be enforced through human resource management. This could take the form of training and education designed to make employees aware of what is acceptable and unacceptable practice and behaviours in the workplace. In this sense, professional conduct and ethical duties of individuals should be applied in the wider public interests and not only confined to the interests of clients and employers. Accordingly, construction professionals must not only take into account the interests of their clients and stakeholders, but the wider communities and the environment as well when taking actions and making decisions.

Adversarial attitudes in the construction industry have in the past affected relationships, behaviours, culture and trust in the construction industry. A major contributor in improving cultures with the construction industry has been professional ethics which defines rules of conduct. Notwithstanding these rules, ethical dilemmas and issues still frequently arise especially in the precontract stages and particularly during the tender processes for projects. With the award of construction contracts normally being synonymous with large sums of money and profits for the successful bidder, it is important to adopt best practices tendering procedures, compliant with the financial regulations of the awarding organisation. Any shortcomings in this regard could open the awarding body up for challenges from unsuccessful bidders especially where the awarding body is a public sector body. For this reason, as part of the tender evaluation processes there is a need to document each stage of the process, including the interviews and clarification sessions and treat all tender submissions equally. This documentation process is important as ethical dilemmas and issues around tender processes have in the past sometimes been challenged and escalated into court proceedings by unsuccessful bidders.

Most organisations normally have rules of governance and declaration procedures to cover corporate hospitality and gifts. However, there may be instances where the self-deception of an individual eludes themselves into the belief that a gift or corporate hospitality will not affect their judgement or impartiality in a particular case. To address this potential dilemma, most organisations have clear guidelines on giving and, perhaps more commonly, receiving gifts and hospitality for the benefit of their employees. In addition, some organisations have strict financial regulations which set down requirements for declaring or rejecting gifts and hospitality over a designated limit. The acceptability of the gift or corporate hospitality will depend very much on its value, but for modest infrequent gifts would normally be very much regarded as a 'token gesture' of goodwill. Unfortunately, there is no universal guidance on what is acceptable or unacceptable in the area of when gifts and hospitality become bribery. However, in deciding whether a gift should be accepted or rejected one needs to consider whether the offering could be considered to be excessive, be regarded as a norm for the business being conducted or whether it could influence unethical conduct or behaviour.

One of the main dilemmas for the community of professional practitioners and professional institutions is related to self-regulation and autonomy. This dilemma is accentuated by the different roles and responsibilities for the industry where diversity creates an environment of competing interest and conflicting ethical practices and standards. In addition, unethical behaviours and practices have in the past been attributed to the self-deception of individuals. In such cases, the self-deceiver's perception of things may encourage and enable them to act in immoral and unethical ways and deluding themselves that they are doing anything wrong in the process. Furthermore, there are scenarios where individuals are unaware that they have acted unethically and where their practices and decision-making can be open to interpretation about what is right or wrong. Such dilemmas if unchecked could provide the platform for accountability, performance and quality to be compromised, to the detriment of clients and the general public. To address this conundrum, it is important for organisations and institutions to have robust regulation and governance of professional ethics with clear guidelines and rules. This could include whistle-blowing policies which have in the past addressed cases related to health and safety breaches, theft, bullying, cover ups, corruption and bribery. Such governance and regulatory policies and procedure are designed to give staff the ability to report such practices and therein provide a more transparent workplace, without the fears of harassment or victimisation.

The next chapter of the book will lead nicely on from this and identify and discuss the various types and examples of unethical conduct and corruption.

References

Abdul-Rahman, H., Wang, C., and Yap, X.W. (2010). How professional ethics impact construction quality: Perception and evidence in a fast-developing economy. *Scientific Research and Essays* 5 (23): 3742–3749.

Ahrens, C. (2004). Ethics in the built environment: a challenge for European universities. The Socrates project. *Proceedings of the 2004 American Society for Engineering Education Annual Conference and Exposition*. 5281–5289.

C.MI. (2010). Code of professional management practice. Retrieved from www.managers. org.uk/code/view-code-conduct [Accessed November 24th, 2019].

Challender, J., Farrell, P., and McDermott, P. (2019). *Building Collaborative Trust in Construction Procurement Strategies*. Oxford: Wiley Blackwell.

Crryside, G. and Kaler, J. (1993). *An Introduction to Business Ethics*, New Edition. London: Centage Learning EMEA.

Fewings, P. (2009). *Ethics for the Built Environment*. London: Routledge.

Hamzah, A.R., Chen, W., and Wen Yap, X. (2010). Date How professional ethics impact construction quality: Perception and evidence in a fast-developing economy. *Scientific Research and Essays* 5 (23): 3742–3749.

IEP (Internet Encyclopaedia of Philosophy) (2010). Ethics and self-deception. Retrieved from http://www.iep.utm.edu [Accessed February 21st, 2013].

Latham, M. (1994). *Constructing the Team*. London: The Stationery Office.

Liu, A.M.M., Fellows, R., and Nag, J. (2004). Surveyors' perspectives on ethics in organisational culture. *Engineering, Construction and Architectural Management* 11 (6): 438–449.

Parson, E. (2005). The construction industry's ethical dilemma. Retrieved from https:// www.ecmweb.com/content/article/20893044/the-construction-industrys-ethical-dilemma [Accessed November 11th, 2010].

Sutherland, E.H. (1983). *White Collar Crime, the Uncut Version*. Binghamton, NY: Yale university Press.

Uff, J.P. (2003). Duties at the legal fringe: Ethics in construction law. *Society of Construction Law*, Retrieved from www.scl.org.uk [Accessed October 2010].

Vee, C. and Skitmore, M. (2003). Professional ethics in the construction industry. *Engineering Construction and Architectural Management Journal* 10 (2): 117–127.

Walker, A. (2009). *Project Management in Construction*, Fifth ed. Oxford: Blackwell Publishing.

5

Types and Examples of Unethical Conduct and Corruption

The educated citizen has an obligation to uphold the law. This is the obligation of every citizen in a free and peaceful society, but the educated citizen has a special responsibility by the virtue of their greater understanding. For whether they have ever studied history or current evets, ethics or civics, the rules of a profession or the tools of the trade, they know that only a respect for the law makes it possible for free people to dwell together in peace and progress.

John F. Kennedy

5.1 Introduction

The above statement from the former President of the United States of America typifies the degree to which upholding the law and not breaching legal or ethical standards is so important for the societies we live and work.

This chapter of the book will identify the various types of unethical and illegal practices from misuse of power to serious fraud. It will articulate what constitutes each type of unethical and illegal practice and how they can occur in a business context. It will provide a list as part of a toolkit and practical guide for what to do and what not to do for ethical compliance. Furthermore, it will explain how construction professionals can decipher what represents unethical conduct and practices in an attempt to regulate their behaviours and actions and those of others. In this regard, as a practical guide it will offer both hypothetical and real-life examples of situations and scenarios where breaches of ethical standards and illegal activity has ensued. Furthermore, it will identify and discuss the damage that can be caused, both at an individual and organisational level, in such cases as a deterrent to bad practices.

The chapter will discuss how the type of unethical or illegal behaviours could vary between different roles and responsibilities within the construction industry from clients down to subcontractors and suppliers in the order of hierarchy. Finally, the measures that can be adopted to reduce unethical and illegal practices across the world will be highlighted with reference to some initiatives that have been spearheaded in the developing world.

Professional Ethics in Construction and Engineering, First Edition. Jason Challender.
© 2022 John Wiley & Sons Ltd. Published 2022 by John Wiley & Sons Ltd.

5.2   How Should Construction Professionals Recognise Unethical Conduct and Practices?

There is a major issue and potential problem relating to the question of what constitutes unethical behaviour in practice. As previously highlighted in Chapter 3 there is no universal theory of ethics with different cultures existing within the construction industry and this creates problems and dilemmas in what is ethical or non-ethical (Liu et al. 2004). Clearly this reinforces the need for construction professionals to have a consistent approach to ethics which can be applied across the whole industry. Liu et al. (2004) explained, however, that this notion of achieving consistency is linked to the different cultures which exist within the construction industry and this may make boundaries between ethical and non-ethical behaviour become blurred at times. A practical example of this could include the boundary between receiving a seasonal gift as a polite gesture and what is deemed to constitute an act of bribery to influence an award of a contract for instance and this was covered in the last chapter. Vee and Skitmore (2003) attempted to address this potential grey area and offered clarity on the boundary between gifts and bribery. They concluded that gift-giving transfers become an illegal act of bribery when they compromise relationships between the gift giver and receiver and favour the interests of the gift giver. This is an important aspect for construction clients, especially at tender stages when bidders may offer them gifts or invitations to corporate functions, to gain competitive advantage over their competitors. It is normal for construction professionals to have to sign up to anti-bribery legislation and declare gifts to avoid accusations of impropriety in such cases. Other forms of unethical behaviour could include breach of confidence, conflict of interest, fraudulent practices, deceit and trickery. What constitutes unethical behaviour in practice is not always black and white and can come in many shades of grey owing to interpretation difficulties. For instance, less obvious forms of unethical behaviour could include presenting unrealistic promises, exaggerating expertise, concealing design and construction errors or overcharging (Vee and Skitmore 2003).

5.3   Examples of Acts of Unethical Behaviour and Corruption

It has been argued that the construction industry provides a 'perfect environment for ethical dilemmas, with its low-price mentality, fierce competition and paper-thin margins'. Transparency International which represents a global coalition against corruption found that 10% on the total global construction expenditure was lost to some form of corruption in 2004, which is staggering and further depletes the reputation of the industry. There have been growing reports over the years that unethical behaviours and practices are still in some areas of the construction industry are still common, and this is taking a toll on public confidence and trust. FMI Capital Advisors (2004) undertook a survey which concluded that 63% of construction professionals felt that unethical conduct was still a problem for the industry. This clearly demonstrates that some improvement measures are required.

According to Parson (2005) unethical behaviours have received negative press coverage and have had a detrimental reputational effect on those organisations involved.

Financial irregularities, scandals, blacklisting and what could the press have referred to as 'dirty tricks' have brought the UK construction industry into disrepute in recent years. Many cases of financial impropriety, false reporting and fraud have entered into the courtroom sometimes resulting in fines, and in extreme cases imprisonment.

In the construction industry, ethical behaviour is measured by the extent by which integrity and trustworthiness is inherent within business conducted by individuals and organisations (Parson 2005). It is particularly complicated in the sector when one considers the different and sometimes conflicting relationships between the project owners (clients), design consultants, main contractors, subcontractor and suppliers. From 'The Survey of Construction Industry Ethical Practices' conducted by FMI Capital Advisors on a large variety of construction stakeholders, the most frequent and reported issues of unethical behaviours related to such activities as overcharging, unfair claims for additional works and bid shopping. The extent of each of these reported practices varied between stakeholder groups. For instance, clients reported their main concern as being overzealous' claims conscious' main contractors whereas subcontractors reported their main issue as late payment, even when main contractors have been paid on time. Main contractors reported that bid-shopping, the practice of clients divulging solicited bids as leverage to encourage them to reduce their lower their prices, was the most concerning for them. According to Parson (2005), this illustrates that the nature and importance of unethical practices changes between the different stakeholders. The other complication revealed by the survey is that the degree to which respondents felt one practice to be unethical is dependent on the type of stakeholder. For instance, the majority of clients felt that the practice of bid-shopping was not unethical but good business practice, and the same was true when main contactors responded to them being 'claims conscious' on projects. Accordingly, this demonstrates that the level of acceptability for certain practices lie in the eyes of the beholder, which will vary between the different parties to the construction process. Subcontractors, who are conscious that they are relatively low down on the food chain sometimes feel that they are the ones who bear the brunt of financial burdens to cut costs on projects, which is commonly referred to as 'value engineering'.

A lot of gamesmanship on construction contracts occurs because the different players have varying degrees of understanding of what constitutes the regulations and rules, especially where tendering and procurement practices are concerned. Accordingly, one party may consider one practice to be acceptable and ethical, whereas another may find it totally unacceptable and unethical. This presents an ethical dilemma for the industry especially as there are no universal standards which offer clarity in this regard. For instance, in some countries reverse bidding, which could be compared with auction sites in reverse, is viewed as acceptable practice, especially amongst construction clients. These individuals might consider that such bidding makes good business sense in securing the best price for the work, but others who feel this is an immoral way to procure tendering processes. Some have argued that reverse bidding could be regarded as ethical as long as the rules were clearly set down to all bidding contractors; however, some would challenge this on the grounds of fairness and transparency.

Another example of questionable behaviour is over-bidding whereupon main contractors and subcontractor charge their clients more than the value of the works they have undertaken, especially at early stages of contracts. This is sometimes facilitated by them 'front-end loading' their financial valuations to reflect being paid more at the start

Table 5.1 Category and frequency of unethical conduct (adapted from Vee and Skitmore 2003). Percentages relate to how frequent respondents had witness that particular category of unethical conduct.

Category of Unethical Conduct	Survey Percentage
Negligence	67%
Unfair conduct	81%
Fraud	35%
Violation of environmental ethics	20%
Bribery	26%
Conflict of interest	48%
Collusive tendering	44%
Confidentiality and propriety breach	32%

than the end of construction periods. Some contractors would argue that this is a necessary and acceptable practice to maintain their cash flow, especially where clients pay them late and deduct monies as retentions in some cases. This may have some legitimacy as most cases of insolvency, especially those smaller construction companies, have been attributed to late payments by their clients.

From a survey carried out by Vee and Skitmore (2003) on construction professionals in Australia, all those interviewed had witnessed or been privy to some degree of unethical behaviours in the past. Such behaviours and practices from the survey are reported in Table 5.1 with percentages relating to how frequent respondents had witness that particular category of unethical conduct.

Clearly the above data is quite alarming when considering the frequency percentages that respondents had witnessed the various forms of unethical conduct in the construction industry.

One attempt to address unethical behaviour in this way comes from The Global Infrastructure Act Anti-Corruption Centre, which published a guide with examples of corruption in the infrastructure sector to assist practitioners. It identified the different criminal acts of fraud including collusion, deception, bribery, cartels, extortion, offences at pre-qualification and tender and dispute resolution (Stansbury 2008). In addition, a survey was conducted by Abdul-Rahman et al. (2007) which looked at ethical problems on building projects in Malaysia and some of the examples of unethical and illegal practices are illustrated in Figure 5.1.

5.4 Misuse of Power

Other less obvious examples of unethical behaviour could emanate from the misuse of power within organisations. This could relate to a manager making unfair or unreasonable demands on one of their subordinates or displaying intimidating behaviour or undermining their staff. In some organisations power can come from an

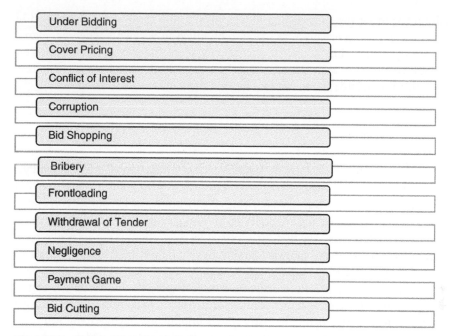

Figure 5.1 Examples of unethical and illegal practices on building projects in Malaysia. Based on Abdul-Rahman et al., 2007.

'illegitimate' form (Walker 2009). This type of power exists outside the formal (or legitimate) authority structure such that the individual exercising power imposes their will on others in the organisation. An example could be an individual who has influence on a senior manager sometimes coined as having 'the ear of the boss'. In such cases, this individual could strongly influence the senior manager's own opinions or actions against another member of staff which can be extremely damaging. This can sometimes occur where there is a relationship or marriage between the individual exerting illegitimate power and the senior manager. Such power can be utilised from an unethical standpoint to further the objectives of the individual rather than those of the organisation. For this reason and coupled with fact that the power is gained outside of the formal line management structure, it can exert a feeling of disquiet in a business context. According to Walker (2009) another category of power that could be regarded as unethical could be 'Coercive' power and is associated with bullying and controlling.

Considering the above, it is perhaps not surprising that the reputation of the construction industry has been tarnished over many years, which has culminated in a general lack of trust and confidence (Challender 2019). This has been a worrying trend and measures are required to address this problem. In instances where ethics have not been maintained this frequently leads to disputes between construction professionals. Some have argued that new players in the construction industry could be emerging with a general lack of appreciation of construction ethics with greed as their main driver. Their conduct in this way can not only affect project outcomes but their ability to manage such aspects as health and safety at work and prevent accidents on site. This is especially relevant for the construction industry given that those people employed in construction contracting are twice as likely to be killed in accidents when compared with the average for all other industries. Accordingly, many individuals have lobbied for the industry to undertake reforms to root out those organisations and people who have breached ethical standards and remove them from the market.

5.5 Corruption

Corruption can be defined as a form of dishonesty or criminal offence undertaken by a person or organisation entrusted with a position of authority, to acquire illicit benefit or abuse power for their own private gain (Fewings 2009). Strategies to counter corruption are often summarised under the umbrella term 'anti-corruption'. Corruption can be grouped into two main areas. The first, 'administrative corruption', is the corruption and bribing of officials, to be treated more favourably. Such officials could include clients for awarding of contracts or local authority representative for receipt of planning permissions or licenses. Moreover, state corruption is associated with individuals trying to change laws or avert the course of justice and can also apply to political figures who act in a biased way to support a particular party who is bribing or lobbying them. Although corruption in not endemic in the UK, there are significant problems which need to be addressed to understand where, why and how it is happening in order to address it. Reasons for corruption are varied but especially in undeveloped economies could be related to underprivileged and underpaid government officials who see acts of bribery as a means to bolster their incomes. However, corruption can ensue in all economies, rich or poor, for reasons of greed and desires linked to furthering their own private interests.

Corruption can be divided into four main categories and these are extortion, bribery, concealment and fraud. Extortion is where a payment is demanded through means of blackmail and where those affected are forced to comply with demands or face the consequences. An example could involve someone gaining personal information on another and uses this information to threaten to disclose this unless a cash payment is given. Bribery is associated with giving or receiving a reward in order to facilitate favours from an individual or organisation. Bribery normally transcends into fraud through false accounting or falsifying documentation for a particular purpose. Tax evasion and wrongfully approving sub-standard work for personal financial gain is treated in the courts as fraudulent activity. Concealment can be associated with corruption and normally involves an agent who facilitates bribe payments. This practice sometimes involves subsidiary companies and transferring and concealing money in offshore accounts for tax evasion purposes. Corruption can take the form of corrupt power where normal competition rules are breached and distorted by a few influential and privileged companies. These can take the form of cartels and serve to increase prices to the detriment of their clients and competitors. These practices can commonly be associated with procurement of building projects at tendering stages and thereby requires robust governance, close scrutiny, reporting and audit to counter such opportunities.

The principle of fair transaction and competition is placed at the heart of ethics and ethical behaviours in a concerted attempt to eliminate corruption. Organisations seek to become transparent in their business dealings to avoid the risk of accusations of unfair or illegal dealings, and therein improve their trading reputation. In some cases, the degree of transparency has to be balanced against the need to maintain client confidentiality which is also a professional responsibility. For regulation of activities there are many areas of business where close scrutiny is called for. These include audit, financial reporting, procurement and remuneration. The concept of transparency is a critical tool in avoiding corruption practices but relies on the principle of eliminating unfair advantage.

Transparency International is a global civil society organisation leading the fight against corruption, and it promotes anti-corruption measures by working in close collaboration with construction participants worldwide. It concurs that it is imperative that all participants in construction projects co-operate in the development and implementation of effective anti-corruption actions in order to eliminate corruption and to address both the supply and demand sides of corruption. The Global Infrastructure Anti-Corruption Centre, (GIACC 2008) has provided 47 examples of construction related activities that are construed as being corrupt and related to the criminal offences of collusion, extortion, fraud, deception, bribery, cartels or similar offences. These are included in Figure 5.2 and are categories into the three categories, namely Pre-qualification and Tendering, Project Execution and Dispute Resolution.

Some hypothetical examples in each of these three categories applied to the construction industry are provided below.

Hypothetical example of corruption (Example 1). Pre-qualification and tendering; Clients obtaining a quotation only for the purpose of price comparison

A client intends to enter into a contract with a preferred contractor for whom he has had previous widespread experience on many projects. The client wants to make sure that the price he pays is in keeping with the market price and certainly no more. In this pursuit, he asks two other contractors to provide quotations for the same job, knowing full well that he has no intentions of employing them. This involves them in time, effort and expense in visiting the site, assessing the works and preparing detailed tenders for the work. The price of one of the two contractors is lower than the preferred contractor, so the client discloses the lowest tender to their preferred contractor and asks them to retender and match the other contractor's price. The favoured contractor retenders on this basis and wins the project, in clear breach of the JCT Codes of Practice for Single Selective Tendering.

Author's verdict

Whilst the above example is not be illegal, it is highly unethical as it has given the other tendering contractors false hopes of winning the project through submitting successful bids and in the process wasted their time, and possibly diverted them from tendering and winning other projects. This can have reputational repercussions for

(Continued)

clients who adopt these practices which could culminate in future contractors refusing to participate in any further tender opportunities. It could also raise questions of why the client in this case would prefer to use the favoured contractor. Some might raise concerns that the client is receiving gifts or bribes from the preferred contractor as a 'kick back' or incentive for awarding contracts to them. In the public sector this type of practice would normally breach organisational governance encompassed in financial rules and regulations. In addition, on those projects over a threshold public sector projects in the European Union have to strictly comply with OJEU regulations and processes. These are designed to ensure that restrictions are not placed on those contractors who can express an interest in being considered for tender, and through robust stages and processes that fair and transparency practices are upheld.

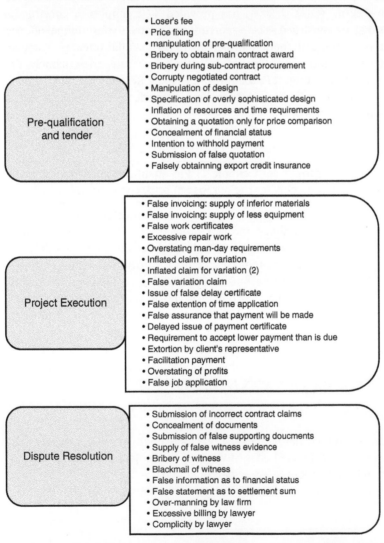

Pre-qualification and tender
- Loser's fee
- Price fixing
- manipulation of pre-qualification
- Bribery to obtain main contract award
- Bribery during sub-contract procurement
- Corrupty negotiated contract
- Manipulation of design
- Specification of overly sophisticated design
- Inflation of resources and time requirements
- Obtaining a quotation only for price comparison
- Concealment of financial status
- Intention to withhold payment
- Submission of false quotation
- Falsely obtainning export credit insurance

Project Execution
- False invoicing: supply of inferior materials
- False invoicing: supply of less equipment
- False work certificates
- Excessive repair work
- Overstating man-day requirements
- Inflated claim for variation
- Inflated claim for variation (2)
- False variation claim
- Issue of false delay certificate
- False extention of time application
- False assurance that payment will be made
- Delayed issue of payment certificate
- Requirement to accept lower payment than is due
- Extortion by client's representative
- Facilitation payment
- Overstating of profits
- False job application

Dispute Resolution
- Submission of incorrect contract claims
- Concealment of documents
- Submission of false supporting doucments
- Supply of false witness evidence
- Bribery of witness
- Blackmail of witness
- False information as to financial status
- False statement as to settlement sum
- Over-manning by law firm
- Excessive billing by lawyer
- Complicity by lawyer

Figure 5.2 Examples of corrupt construction-related activities (adapted from GIACC 2008).

Hypothetical example of corruption (Example 2). Pre-qualification and tendering; contractor price fixing

A group of contractors who frequently tender for works in the same market, enter into a secret arrangement wherein they share work between themselves. Unbeknown to the client they secretly decide which of them will win each tender, despite being seen to be competitively and tendering for each project. The contractor who is chosen to win a tender will notify the other tenderers as to their tender price in order that the other submit higher bids. In this way they secure that the chosen contractor is the successful bidder and therein awarded the project. The successful contractor has the clear opportunity to place a higher bid than they would otherwise have submitted if they were competing on a level playing field in the full knowledge that such a bid will be guaranteed to be successful. If sufficient contracts are placed each of the same contractors will have the opportunity to become the successful bidders and, in this way, they guarantee themselves a share of the projects awarded and at a higher price than would be determined by market forces. This effectively means that the client pays more than they would under open competition expect to pay.

Author's verdict

In the above example, the tendering contractors are disadvantaging the client who is unaware that price fixing arrangements are in place, and therein the client is paying more for a particular project as the rules of competition have been breached. One could argue that the costs of tendering for the contractors is reduced also as little if any due diligence has to be undertaken when constructing their bids. They simply can set a price which reflects an increased profit, knowing full well that they will be awarded the contract.

Hypothetical example of corruption (Example 3). Project execution; inflated claim for variations

A contractor is instructed by a contract administrator, acting for a client, to carry out some variation works to the contract. In consideration of variations, the building contractor claims for additional payment and depending on the scope, possibly to an extension of time to the contract programme. This extension of time carries with it a further additional cost and time allowance. The contractor purposefully uses the variation to secure a commercial advantage over their clients by exaggerating the claim and the overall effect on the contract. This could be by false accounting of materials, manpower, equipment and time to undertake the changes or additional work. In addition to claims for financial recompense for additional works, they claim for a financial impact on them through disruption. Furthermore, an extension of time is requested by the contractor to push the completion date back, coupled with a further claim for additional monies for overheads, profits and preliminaries for each week the contract is extended. This is frequently referred to as 'prolongation costs' on the contract.

(Continued)

(Continued)

Author's verdict

The above example could lead to disproportionate and unfair and disproportional additional costs and time extensions being incurred by the client for the variation on the contract. They could be paying more than they should for the changes or additional work that they have instructed under the contract. In some cases, contractors have claimed for an extension of time brought about by relatively insignificant events for which would have limited or no effect on the contract. This is a frequent source of how disputes are generated on projects, which sometimes cannot be resolved by the contractual parties and are referred to adjudication for judgement.

Hypothetical example of corruption (Example 4). Project execution; concealing defects

A roofing subcontractor installs a waterproof covering on a building. The covering is accidentally damaged while being laid to the extent that the perforated membrane could cause future water penetration into the building. The roof works require to be certified by the clerk of works who is responsible for inspecting the roof before it is covered over. The membrane is clearly substandard as a result of the damage and should be rejected and replaced owing to the perforations. The clerk of works is offered a financial reward for certifying the roof covering as acceptable and in conformity with roofing standards. The work is certified, with both parties knowing full well that the work is defective and prone to future leakage. The roofing subcontractor issues the roofing certificate to the main contractor as a guarantee that the works has been carried out satisfactorily and obtains full payment for the defective roofing works. Neither the subcontractor or the clerk of works discloses to the main contractor or anyone else that the roof membrane is perforated and requires replacement.

Author's verdict

In the above example, the client is ultimately paying for a defective roof which at some point in the future will have to be repaired or replaced as significant cost. The roofing subcontractor has concealed a major defect to avoid having to incur the cost of repair or replacement of the roof covering. In addition, the clerk of works has accepted a bribe to certify the roof as satisfactory knowing full well that it is defective. Both the clerk of works and subcontractor have acted fraudulently for financial gain and the client building owner is left with a defect that will require rectification at a later stage.

Hypothetical example of corruption (Example 5). Dispute resolution; submission of false supporting documents

In a contract claim, or dispute resolution proceedings, a claimant submits the following supporting documents as accurate and genuine when he or she knows they are false, or does not believe them to be true, or at best cannot be sure of their accuracy. Timesheets are submitted by a contractor which are not accurate or genuine and which have been produced to falsely show that equipment and labour was used to procure a particular item of work, when in fact it was not.

Work records which are also not genuine or accurate, have been created to overstate or describe incorrectly materials and equipment supplied.

Programmes and timelines which proport correctly to state events and dates, have been deliberately amended to attribute delay falsely to a stated cause. Furthermore, cost records which incorrectly state the cost of items, or include item of work which were not in fact provided.

Photographs were created to falsely show that an activity occurred at a certain time, or location, when in fact it did not.

Author's verdict

In the above example, the claimant is quite clearly presenting an inaccurate, false and some would say fraudulent position in order to serve their own purposes. Such purposes could revolve around financial claims, extension of time, liability for health and safety breaches, including accidents or some other event where a dispute with the other contracting party has occurred. If the other party is able to prove that falsification of documentation has taken place this could become the basis of a criminal proceeding against the perpetrator.

Hypothetical example of corruption (Example 6). Dispute resolution; false statement as to settlement sum

A main contractor reaches an agreed final account sum with their client for a particular project. The final account reflects a payment in full for all subcontractor work packages. Notwithstanding the subcontractor payment allowance, the main contractor falsely reports to one of the subcontractors that they have received a lower amount for the subcontracted work from the client than they have actually received. They have claimed that the client has reduced monies owed owing to defects or snags with the subcontract work completed. The main contractor uses this position to agree a reduced settlement with that particular subcontractor for the work they have carried out.

(Continued)

(Continued)

Author's verdict

In the above example the main contractor has misled the subcontractor into believing that they have not been paid in full for the subcontractor work and could have falsely articulated many different reasons for this, e.g. poor workmanship. This has disadvantaged the subcontractor and led to them not being paid the correct amount, possibly leading to them making a loss on a particular contract. Some smaller subcontractors may find this financial impact too much to manage in terms of cash flow and in extreme cases could cause the subcontractor to become insolvent.

Whilst the above cases are hypothetical, there have been many reported real-life examples of fraudulent activity in the construction sector. Where this has occurred, it has been very damaging for the reputation of those organisations, with them being found guilty of breaching ethical standards and committed criminal offences. The case of construction and professional services company Sweett Group PLC in February 2016 was one such example on a global scale, where they were sentenced and ordered to pay £2.25m after a bribery conviction. The case involved their failure to prevent an act of bribery in the United Arab Emirates to secure the award of a contract for the building of a hotel in Abu Dhabi. The Serious Fraud Office (SFO) uncovered that a subsidiary company, Cyril Sweet International Limited, had made corrupt payments to a Real Estate and Investment arm of Al Ain Ahlia Insurance Company (AAAI). The Director of the Serious Fraud Office David Green QC articulated after the case that:

> *Acts of bribery by UK companies significantly damage the country's commercial reputation. This conviction and punishment, the SFO's first under Section 7 of the Bribery Act, sends a strong message that UK companies must take full responsibility for the actions of their employees and in the commercial activities act in accordance with the law.*

This caused worldwide negative for the organisations involved and the construction industry at large. Sweet Group PLC were ordered to write to all their existing clients to provide a full and comprehensive account of the conviction which led to extremely bad press for them. The full details of this case will be covered later in this chapter. This potentially could have caused them to become insolvent, due to reputation damage and significant decrease in work as a consequence.

5.6 Fraud

According to Fewings (2009) fraud can pose one of the biggest risks to organisations. Fraud can take many different forms but most prevalent in the UK is cybercrime. In the past, there have been instances where emails have been received that have stolen the identities of organisations. Where clients have large construction projects on site, it can be sometimes a magnet for acts of fraud to be enacted. This could start with a perpetrator calling the client and pertaining to be the contractor, advising that their bank details have changed and requesting that the client make the necessary changes on their finance system. Procedures should be in place not to comply with bank change requests of this nature but nevertheless a financial controller should not make the changes without adherence to special controls and procedures. Coupled with the change of bank details, frauds of this nature involve the raising of a bogus invoice claiming to be from the contractor, after which bank transfers are made to them and the monies fraudulently channeled through to the criminal organisation. In such circumstances money transfers normally are wired oversees and the Serious Fraud Squad in the UK find it extremely difficult to identify the perpetrators who have taken on a false identity. Retrieval of any funds transferred can be even harder and insurers do not normally cover organisations for such loses. This can leave a company in a poor financial position as a consequence of the financial loss. In a construction context consider the hypothetical example below:

Hypothetical example of fraud involving cybercrime

A large multi-national main contractor is employed by a college to construct a new building for them to the value of £20m. Halfway through the contract duration, the financial accounts department of the college receive an emailed letter from what proports to be the main contractor. It appears genuine and authentic as it is on letterheaded paper of the main contractor and all the right VAT and company information details. The letter advises the College that the contractor have changed banks and accordingly they request that all future monthly payments are changed over to the new details which they have detailed in the letter. This letter is very shortly followed up by a phone call again claiming to be from the main contractor's financial representative to the financial accounts department, enquiring whether they received the letter and asking for the next due payment in excess of £1m to be paid into their new bank account. The financial officer who deals with both the letter and the phone call is relatively inexperienced and does not seek to escalate the change request to superiors. Couple with this, there are no procedures in place for dealing with requests such as these and no approval protocols, especially around checks to assess the authenticity of the request. Accordingly, the financial officer makes the change on the system and on receipt of an invoice the monies are paid into the new bank details. It later emerges when the main contractor contacts the college to complain about late payment that the contractor has not changed banks and the college become aware that they have been subjected to a fraudulent activity with payment made to a criminal organisation. The college is not insured for such an event as is classed as negligence on their part and the funds paid cannot be traced by the Fraud Office. The College loses £1m in this instance and owing to the matter receiving press coverage, reputational damage has been caused to the institution.

(Continued)

(Continued)

Author's verdict
Albeit this is a hypothetical case, scenarios such as this have taken place many times in the past and all companies and institutions should have robust financial regulations in place for preventing such fraudulent practices to avoid criminal activity and the financial aftermath of such eventualities.

5.7 Bribery

One of the most common forms of unethical conduct is bribery which is described as:

> *Offering some goods or services for money to an appropriate person for the purpose of securing a privileged and favourable a privileged and favourable consideration of one's product or corporate project.*
>
> Vee and Skitmore (2003)

Bribery relates to a conscious effort to influence another person or organisation with a gift or payment that generates an unfair advantage. It normally involves moving from a position of providing an acceptable payment for goods or services to an attempt to distort an outcome to generate an advantage (Fewings 2009). According to Usman et al. (2012) there are two types of bribery; 'Active' and 'Passive'. Active bribery relates to the promise or offer of any financial or other advantage, directly or indirectly with the intention to induce or reward a person for the improper performance of a function. Passive bribery relates to the request, acceptance or agreement to receive any financial or other advantage, directly or indirectly, as a reward intending that a relevant function will be performed improperly.

The offering of bribes could be considered as inducements to someone in a position of authority and trust in getting them to do something for the person offering the bribes, which they would not otherwise be entitled. The Bribery Act 2010 makes it an offence to accept or give or receive in kind inducements to receive favours. Notwithstanding this premise, supported by the Law, Vee and Skitmore (2003) opined that this is an area where there are seldom scenarios of 'black and white' but more so lots of cases where grey areas exist. This is related to the fact that persons in authority frequently receive what could be described as 'gifts' in the form of 'hospitality'. The question therefore arises when does 'gift giving' become bribery. Some would argue that this depends on the extent and degree of such gifts or hospitality (Vee and Skitmore 2003). By way of examples, many organisations have policies on gifts and hospitality and have procedures in place which place limits on the value of such offerings. Other companies may have rules which places requirements on staff to declare anything that they receive from third parties. Notwithstanding, these management systems and procedures for dealing with gifts and hospitality, do rely on the interpretation and the discretion of the individuals receiving them. The aforementioned grey area that surrounds this subject matter, has forced some organisations to have a zero tolerance on any form of hospitality or receipt of gifts. This may be regarded as an extreme measure especially where with disciplinary measures for those who choose not to abide by the organisational mandate. However, some would argue that it is the only way to ensure that abuse of such offerings does not take place.

According to Johnson (1991), the following criteria has to apply for gifts or hospitality to be deemed the illegal act of bribery:

1) The gift or hospitality must be of a non-token nature that it considered reasonable to provide the giver with privileged status or
2) The beneficiary of the gift may consciously or otherwise to favour the interests of the gift giver.

Some organisations have allowed gifts to be received and gift making to their clients as long as the above two conditions have not been breached.

Bribery laws and associated ethical practices apply to all organisations and their employees. It also applies to senior managers who 'turn a blind eye', suppliers and potential suppliers. There are many different risk areas where bribery can frequently take place and where there are temptations for potential bribery in the construction industry. Procurement is one of these key risk areas especially when large multi-million-pound contracts are being awarded. There have been many international cases that have occurred and been widely publicised. The Sweett Group PLC is one such case where fines of £2.25m were imposed after a conviction under the Bribery Act. The details of the case are detailed below:

Example of a case involving bribery: The Sweett Group PLC

In 2012, Sweett's Cypriot subsidiary secured a contract with the Al Ain Ahia Insurance Company (AAIC) to manage the construction of a hotel in the United Arab Emirates (UAE). Separately, Sweett entered into a second contract with North Property Management (NPM) for associated 'hospitality services'. One of AAIC's officers, Mr. Al Badie, was the beneficial owner of NPM. While payments were made to NPM under the second contract, there was no record of Sweet receiving any services from NPM. The contract was summarised by the sentencing judge (HHJ Beddow) as a 'vehicle to provide a bung'. On September 2015, Sweett admitted a failure to prevent bribery. Under s7(1) of the Bribery Act 2010 (the Bribery Act), a company is guilt of an offence if a person associated with it bribes another person to obtain business. While under s7(2) there is a statutory defence if a company can show it had in place adequate procedures designed to prevent the bribery occurring, Sweett admitted that it had not.

The actions of a foreign subsidiary carried on outside the UK, in relation to a UAE contract were caught by the Bribery Act. Whilst Sweett's Cypriot subsidiary had committed the offence, the systematic failures of Sweett, as the parent company, to properly supervise its subsidiary made Sweett liable.

(Continued)

(Continued)

Lessons learnt from this case

Whilst this case is quite extreme many have asked what lessons can be learnt from it. From what can be gathered from the case, Sweett's dealings with the Serious Fraud Office (SFO) had become tenuous and the relationship had deteriorated during the period of the investigations. The SFO has reportedly advised Sweett to stop trampling on the evidence. Nonetheless, there are a number of lessons to be drawn.

The importance of self-reporting was one lesson. Sweett had been linked to an earlier bribery involving the construction of a hospital in Casablanca, but Sweett had only reported the payments at a very late stage to the SFO ahead of a *Wall Street Journal* article being published. Furthermore, Sweet only reported on other connected payments as being potentially suspicious when the SFO began formal investigations into Sweett. This late self-reporting was considered bad practice by the SFO. Sweet were also accused of hedging their bets by forming a defence that the payments under investigation were part of a finder's fee arrangement, and to add 'insult to injury' these payments continued to be made for a significant period of the investigation. This case from an ethical perspective also highlighted the unwillingness of Sweett to fully cooperate with the SFO investigation and this included handing over their accounts of witness interviews. This caused further reputational damage to them coupled with a significant fall in their share price.

The Sweett Group PLC is not an isolated case and in recent years there have been many more examples. The individual consequences can have devastating effects on individuals and organisations. These include the following:

- Damage to reputation (personal and organisational)
- Liberty (maximum 10 years prison sentence)
- Financial implications (unlimited fines and compensation orders)
- Inability to participate in public contracts for organisations

According to Global Infrastructure Anti-Corruption Centre (GIACC 2008), in terms of learning lessons from previous examples of bribery, there may be a list of what to do and what not to do and these are contained in Table 5.2.

5.8 Conflicts of Interest

An issue that has a link with corruption is a conflict of interest. According to Fewings (2009) a conflict of interest is '...the conflict where a single person has obligations to more than one party to the contract, or there is a clash of private or commercial interest with a person's public position of trust'. Conflict of interest could be defined as an interest, which if pursued, could keep professionals from meeting one of their obligations. Sometimes conflicts of interest could relate to the right of an employee to refuse to

Table 5.2 Lessons learnt from previous cases of bribery (adapted from Global Infrastructure Anti-Corruption Centre 2008).

Things to Do	Things Not to Do
Be aware and sensible	Be naive
Be transparent	Think you are doing the organisation a favour
Get authorisation	Be prepared to accept anything without checking such as: Regulations Code of conduct Your line manager Independent advice
Apply the 'sniff test'	Do something which you would not want publicising

conduct unethical behaviour. Such refusal could relate to an individual not wanting to partake in an activity that they regard as immoral or one that they believe is fundamentally wrong.

Notwithstanding the above definitions, a conflict of interest could also be defined as a clash of commercial and private interests by individuals who hold a position of authority and trust. It could also be synonymous with an individual who has interests or obligations to two or more parties to a contract. Sometimes large national or global consultancies find themselves in instances where they are representing two different organisations who may be involved in the same project or party to the same contract. In most cases, this may cause the consultancy to withdraw from one party to avoid becoming conflicted. Notwithstanding this more common scenario, it is sometimes commonplace for those same consultancies to accept both commissions and form a 'Chinese wall' between two parts of their organisation to deal separately with the two clients. There are ethical questions as to whether this type of practice is acceptable. What is of paramount importance in such scenarios is that their clients are consulted to make them aware of the possible conflict of interest and give them the powers to withdraw them from acting for a client in such cases. If their client's approval is expressly given, then it is imperative that the Chinese wall maintains a completely independent position for both clients without collusion and even contact in some instances. Notwithstanding the above, most lawyers would not advocate the use of the same consultancy to represent two parties to the same contract. Afterall it could become very complicated and convoluted if there was a dispute between the two contracting parties and the same consultancy representing both parties was required to give evidence in court. The author is not aware of any such cases but with such a known risk, many consultancies should steer clear from such practices just in case to avoid potential reputational damage.

In some cases, client's boundaries have become blurred when working with their supply chain and this can result in a conflict of interest. An example could be a client representative who asks a contractor that they have appointed and worked with on construction projects to price an extension on their own house. This request is not strictly speaking the client seeking a financial concession on this private work and as such they may believe that this is not unethical in any way. Possibly this could be a classic case of an ethical dilemma brought about by self-delusion as discussed in the last chapter. The fact that the client has chosen to use a known contractor who they hold influence over and who is reliant on them for work could be perceived to be morally wrong. The contractor may feel they cannot refuse to provide a quotation for the works but could also believe that they would be expected to undertake the work for a lower price than the market would dictate. This puts both the client and the contractor in a compromised position where favours are given and expected in return. Therein lies the real problem in this area in that, certainly in a business context, people do not normally do things for nothing and there is a reciprocal reward that is sought at a later stage. Using the same example this could involve an obligation for the client to provide the contractor with a future contract award. This notion is captured by the phrase 'you stroke my back and I'll stroke yours'. Accordingly, the safest position is to separate business and private dealings especially when you an individual is working from a position of authority and influence and 'better to be safe than sorry'. Two different examples of conflicts of interest are described below:

Example 1 of a case where there is a conflict of interest as a Board member

An individual who is a member of a university Finance and Resources Board responsible for approving large capital projects, is aware that they have a personal vested interest on one of the projects under board consideration. This relates to a company they own supplying consultancy services to a contractor who has been selected for the project. Furthermore, the Board member concerned will benefit from additional consultancy work if the project is approved by the Board. Despite this, the individual considers that their company will not be directly appointed by the University and under a misguided and possibly conscious effort to accept there is an issue, discounts any notion that there is conflict of interest. Accordingly, they do not declare an interest at the start of the meeting and remain silent on their role in the new project under consideration by the Board. The project is subsequently approved, and the Board remains ambivalent to any personal interests that any of its Board members have in this particular case.

Author's verdict

The above case could be regarded as a blatant attempt by the Board member to gain personal advantage from an approval process that they are directly party to and which they have actively participated within. In some case, it could be considered that they have subconsciously persuaded themselves that an actual conflict does not exist and again self-delusion could be at play in this scenario. The Board member has been unable to remove the strong temptation to remain impartial which could be construed as self-denial of their part. If their personal interests in the project came to light, they could be dismissed from the university Board and possibly suffer reputational damage brought about by claims they have deliberately withheld important information. The most professional and safest position for them would have been to declare an interest and remove themselves from the debate and deliberations on that particular project.

Example 2 of a case where there is a conflict of interest on a planning application

A similar scenario as the above example could present itself where a planning application is being considered by a Council Planning Authority and one of the Planning Committee councillors has not declared an interest. This could relate to the same councillor has a vested interest in a particular application, possibly where their own property is directly affected by a neighbouring development. This would propel them to persuade their fellow committee members to reject the application on planning grounds, when their real motivation is to avoid any detrimental impact on themselves.

Author's verdict

The councillor in question is duty bound to disclose an interest in any planning applications which they consider, and this is clearly designed to avoid them personally benefitting from the decision-making process. This disclosure would have been a condition of joining the Planning Committee. Following strict Council governance regulations and requirements they would have had to commit in writing to declaring an interest in such cases. The fact that the Councillor has deliberately chosen to conceal their private interest in a public process, could amount to a fraudulent action and certainly a breach of the rules and regulations that they signed up to as a member of the planning committee.

5.9 Ethics and Negligence Linked to the Design and Construction of Buildings

Cases of negligence could be unearthed in the whistle-blowing procedures of organisations as discussed previously in Chapter 4. Such procedures are clearly designed to eliminate or reduce bad practices and short cuts that could lead to claims for damages against those organisations. Negligence in this regard could be described as:

> *The failure to exercise that degree of care which, in the circumstances, the law requires for the protection of those interests of other persons which may be injuriously affected by the wants of such care.*

Delbridge et al. (2000)

The above quotation, in the context of negligence within the construction industry could be extremely relevant. Where contractors or designers have compromised the health, safety and well-being of individuals, including members of the public, this could have catastrophic results with potential loss of life. Some examples of catastrophic cases involving accusations of negligence are contained below:

Case study 1: Negligence in design and construction

Figure 5.3 Photograph of Grenfell Tower, London, following the devastating fire in 2017.

One example could be the case of the Grenfell Tower disaster in London in June 2017 (see photograph in Figure 5.3). A fire was started by a malfunctioning fridge-freezer on the fourth floor. It spread rapidly up the building's exterior, bringing fire and smoke to all the residential floors and a death toll of 71 lives. The cause of the spread, at a public enquiry carried out in October 2019 concluded that insufficiently designed combustible cladding caused serious fire to spread throughout the whole building.

The *Daily Mail* (2020) reported that a multi-million-pound company which made the flammable cladding delayed approval of a safer version purely based on cost-saving grounds. US construction giant Arconic supplied polythene panels through a subsidiary company AAP SAS in France. Grenfell survivors and relatives of the bereaved have launched a lawsuit in the US against Arconic, accusing the company of putting lives at risk by supplying an unsafe product for residential use. They claim the company acted out of corporate greed by supplying a product which is banned on US buildings over 12m. US District Judge, Michael Baylson reported that the allegations of cost savings 'presents strong evidence of egregious conduct by Arconic'. He said that Arconic refused to approve a safer cladding from AAP as it was regarded as too expensive. The lawsuit claims that Arconic had a desire to cut corners and save money culminating in a reckless decision to supply the cheaper cladding which would have been £293k less than the fire-retardant version. Arconic's defence revolved around them warning customers that it's cladding represented a substantial fire risk for high-rise buildings, and therein not recommended for such buildings over 12m.

Despite these findings, the question of whom is negligent and therein responsible will no doubt be a matter of contention for many years to come. This dilemma is compounded by the US District Judge in September 2020 dismissing the case and ruling that it would be more efficient if it was tried in the UK. A subsequent appeal against this decision has been brought by the relatives.

Other parties affected by the tragedy, other than the victims and their families, included the thousands of leaseholders who have discovered their buildings contains the same combustible fire panels as those used on Grenfell. Most of these individuals have been presented with large bills associated with overseeing the retrofitting of new cladding to their buildings, which in a lot of cases is unaffordable for them. Such costs have ranged between £40,000 and £115,000 to replace the dangerous cladding, which has prevented them from selling their properties and could make many householders bankrupt. Whilst waiting for these repair works, most residents have been required by their building insurers to employ 'Waking Watches' to safeguard the buildings and residents in the intervening period. In addition, a further amount of up to £2.2bn could be incurred a year between them for stop gap measures and costlier insurance (Daily Mail 2021a). Despite this predicament for many, building firms linked to Britains's unsafe housing crisis have made profits in excess of £15bn since the disaster. Furthermore, the 10 largest construction firms paid their chief Executive Officers an average total salary on £2m in the 2020 financial year (*Daily Mail* 2021a).

Questions that have emerged from this tragedy

With such a devastating and catastrophic tragedy with major loss of life how could this have happened in the UK, given that we are one of the world leading economies in construction?

At an inquiry held in the UK in February 2021, it was revealed that the cladding fitted on the Grenfell Tower burnt 10 times faster than a sister product, but both sold under the same fire safety certificate. Furthermore, it was reported at the same inquiry that the cassette-type panels fitted on Grenfell Tower emitted three times as much smoke and seven times the amount of heat than another rivetted type of cladding panel (*Daily Mail* 2021b). Accordingly, the cassette type performed spectacularly worse than its contemporary, despite both being sold in the UK and no distinction being made between the two. The Managing Director of Arconic, Claude Schmidt, who took several months to agree to give evidence at the enquiry, was forced to admit that the fire classification of the cassette panels was false due to 'incomplete information'. Furthermore, he conceded that he did not seek to understand the fire testing and classification which supported the sale in the UK, as 'it was not a priority for him'. Notwithstanding this undertaking, he refused to accept personal responsibility for the tragedy but did accept that Arconic were responsible for the sale of the cladding to the UK Market on the grounds of incomplete information. Clearly this dereliction of duty and non-acceptance of personal responsibility was no comfort to the many victim's families, and some would say added 'insult to injury' for them in this tragedy.

Is it fair to apportion blame on the designers, main contractors, subcontractor, suppliers and cladding manufacturers of what is known now to be a disaster waiting to happen? Alternatively, should it not be the role of Government to be the instruments of standards and ethics in the UK and ensure that sufficient and appropriate regulation and due diligence is undertaken? This could involve more stringent requirements around certification and regulation of such building components in the future or changes to Part B of the Building Regulations? How can we ensure that ethics in designs and testing of building materials and components is sufficiently embedded in the UK construction industry to avoid a repetition of this awful tragedy?

Case study 2: Negligence in design and construction

Figure 5.4 Photograph of Ronan Point tower block, London, following its partial collapse in 1968.

Another less recent case of negligence in the construction industry related to the 22-storey Ronan Point tower block in Canning Town, east London (see photograph in Figure 5.4). There was a partial collapse of the building in 1968, only two months after it had been constructed. A gas explosion blew out some load-bearing walls, causing the collapse of one entire corner of the building, which killed 4 people and injured 17 others. The spectacular nature of the failure (caused by both design and poor construction) led to a loss of public confidence in high-rise residential buildings, and major changes in UK Building Regulations followed.

The above examples are clearly related to negligence associated with design and construction flaws, but it is interesting to also consider ethics from a construction, health and safety at work perspective. This is particularly pertinent to the UK construction industry where 2019 annual figures from the Health and Safety Executive showed that 30 people were killed on construction sites between April 2018 and March 2019. The leading causes of construction worker deaths were falls from height, which continue to be the biggest cause of death on construction sites despite calls for more safety procedures in previous years. These cases in totality accounted for more than half of the construction death reported in 2019 and were followed by impact from objects and electrocution. An example of a health and safety case breach which transcended into criminal proceeding is detailed below:

> **Case study 3: Negligence related to construction phase health and safety breaches**
>
> A company was fined £3000 in 2018 and ordered to pay £6,500 in costs after a construction worker fell five metres through a roof and subsequently died. It was deemed by the Health and Safety Executive (HSE) following their investigation that the tower scaffold used to access the roof was damaged and had not been correctly erected. The company who erected it, namely Maidstone Studios, had purchased a second-hand tower scaffold with no instruction and without having any competencies to erect such a scaffold. There was no certification, tests or checks made on the scaffolding structure, and no training had been given to the deceased construction worker on how to use the scaffolding. The HSE concluded that this was a senseless loss of life that could have been averted if the work had been planned properly and suitable access equipment provided, correctly placed and erected by those with adequate training.

The above case studies should provide the food for thought on how such blatant examples of negligence and unethical practices, with such obviously devastating consequences, should be avoided in the future. This area will be covered later in the book in Chapter 6 around regulation and governance of ethical standards and expectations to counter unethical behaviour.

5.10 Global Corruption in the Construction Industry

A study by Usman et al. (2012) looked at unethical practices in the management of construction projects in Nigeria. According to them approximately 10% of the value of construction projects undertaken by the government of Nigeria is used to bribe government officials as 'kick backs' for granting work orders. This has had the overall effect of influencing government officials to initiate some projects for the financial benefits they could benefit from in this regard. Whilst some would argue that this is accepted as normal practice in Nigeria it is nevertheless an illegal and highly immoral practice which needs to be outlawed and prevented from occurring. They are benefiting a small minority of influential people at the expense of most of the population and in the process adversely affecting the fortunes of the country.

A further study by Abdul-Rahman et al. (2007) presented findings against frequency of unethical conducts in the Malaysian construction industry. These are illustrated in Table 5.3 and grouped into 11 categories ranging from acts of under bidding to bribery in the extreme. They also classified these categories against a frequency scale, based on Never, Sometimes, Often and Very Often, for each of these practices to ascertain the scale of their occurrence. Their analysis found that the top two ranked most unethical practices, classified as an 'Often' occurrence, related to what could be described as serious unethical practices around under bidding, bribery and corruption. This could be regarded as quite controversial and damaging for the reputation of the construction industry, certainly within a Malaysian context. Moreover, the next three most ranked unethical conducts, based around negligence, claims and payment games were found to have occurred 'Sometimes'.

Each year Transparency International scores countries on how corrupt their public sectors are seen to be. It sends a powerful message and governments have been forced to take notice and act. Behind these numbers is the daily reality for people living in these countries. The index cannot capture the individual frustration of this reality but does capture the informed views of analysts, businesspeople and experts in countries around the world. The Construction Perception Index (CPI) scores and ranks countries and territories based on how corrupt a country's public sector is perceived to be by experts and business executives. It is a composite index, a combination of 13 surveys and assessments of corruption, collected by a variety of reputable institutions. The CPI is the most widely used indicator of corruption worldwide. The top 17 countries in the CPI survey for 2019 are shown in Table 5.4. These countries scored above 75 in their index, with Denmark and New Zealand topping the list of countries with scores of 87 and four others namely Finland, Singapore, Sweden and Switzerland scoring 85 or above. The UK came at rank number 15 with a score of 77. Moreover, the bottom 17 countries in the CPI are shown in Table 5.5. These countries scored a CPI of 20 or less and included those countries that had undergone political and economic turmoil in recent years, mostly resulting from wars and instability.

Table 5.3 Ranking of unethical conduct by construction players (adapted from Abdul-Rahman et al. 2007). Rank No. 1 = Most frequent. Rank No. 11 = Least frequent.

Categories of Unethical Conducts	Rank
Underbidding, bid shopping, bid cutting	1
Bribery, corruption	2
Negligence	3
Front loading, claim game	4
Payment game	5
Unfair and dishonest conduct, fraud	6
Collusion	7
Conflict of interest	8
Change order game	9
Cover pricing, withdrawal of tender	10
Compensation of tendering cost	11

Table 5.4 Top 17 construction perception index (CPI) scores and rankings by country.

Country	Ranking	CPI Score	Number of Sources
Denmark	1	87	8
New Zealand	1	87	8
Finland	3	86	8
Singapore	4	85	9
Sweden	4	85	8
Switzerland	4	85	7
Norway	7	84	7
Netherlands	8	82	8
Germany	9	80	8
Luxemburg	9	80	7
Iceland	11	78	7
Australia	12	77	9
Austria	12	77	8
Canada	12	77	8
United Kingdom	12	77	8
Hong Kong	16	76	8
Belgium	17	75	7

To reduce corruption and restore trust in politics, Transparency International recommends that governments:

- Reinforce checks and balances and promote separation of powers.
- Tackle preferential treatment to ensure budgets and public services are not driven by personal connections or biased towards special interests.
- Control political financing to prevent excessive money and influence in politics.
- Manage conflicts of interest and address 'revolving doors'.
- Regulate lobbying activities by promoting open and meaningful access to decision making.
- Strengthen electoral integrity and prevent and sanction misinformation campaigns.
- Empower citizens and protect activists, whistle-blowers and journalists.

In any country, especially those experiencing a high level of development opportunities, it is imperative to have a high level of ethical standards in construction management, as clearly construction plays a major role in economic development. For this reason, construction professionals need to demonstrate and abide by a high standard of social responsibility and behaviour. This should extend to individual judgement, expertise and accountability with regard to the implementation and management of

Table 5.5 Bottom 17 construction perception index (CPI) scores and rankings by country.

Country	Ranking	CPI Score	Number of Sources
Iraq	197	20	5
Burundi	197	20	6
Congo	199	19	6
Turkmenistan	199	19	5
Democratic Republic of Congo	201	18	9
Guinea Bissau	201	18	6
Haiti	201	18	6
Libya	201	18	5
North Korea	205	17	4
Afghanistan	206	16	5
Equatorial Guinea	206	16	4
Sudan	206	16	7
Venezuela	206	16	8
Yemen	210	15	7
Syria	211	13	5
South Sudan	212	12	5
Somalia	213	9	5

construction projects. In Malaysia, a governing body was established in partnership with stakeholders from industry, namely the Construction Industry Development Board (CIDB). The aim of this board was designed to tackle unethical practices and promote good ethical values in the construction industry, and they have developed the Construction Industry Master Plan (CIMP) as a route map in this regard. The underlying aim of this revolved around improving service quality outcomes and professionalism for the benefit of all stakeholders and communities.

5.11 Effect of Unethical Practices and Corruption

Apart from human values and morality issues, there can be significant damage for organisations in terms of economic loss, lawsuits and settlements from ethical breaches. Costs to businesses from unethical practices and behaviours can also affect employee relationships, damage reputations and reducing employee productivity, creativity and loyalty (Weiss 2003). Within the context of the construction industry, one of the dilemmas and challenges is that firms solely focus on the ethical behaviours of their own organisation. Construction companies normally have large supply chains that have to be part of the business ethics model. For this reason, they sometimes have an ethical and moral responsibility for the behaviour of their suppliers, subcontractors and other stakeholders. This could provide a positive way of influencing other organisations down the chain for reforming and educating the construction industry.

According to Inuwa et al. (2015) there has been little research undertaken in the past relating to the effects of professional ethics on the performance of construction projects. In developing countries especially, there have been examples of corruption severely affecting the quality and in some cases compliance of buildings with current regulations. In some cases, this has resulted in adverse events as show in Figure 5.5.

In extreme events this has resulted in the following outcomes as show in Figure 5.6.

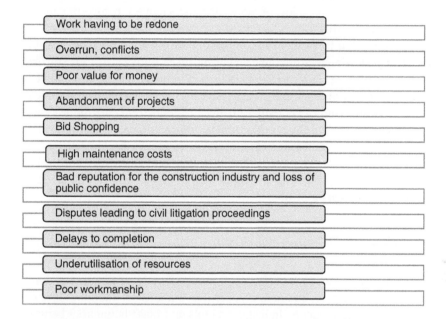

Figure 5.5 Examples of adverse extreme events resulting from corruption.

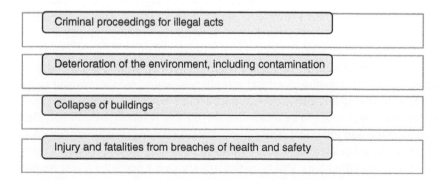

Figure 5.6 Examples of extreme negative events resulting from corruption.

5.12   Remedies for Unethical Behaviours and Corruption

A high level of professionalism is expected of construction professionals in discharging their duties and strictly adhering to ethical standards. However, this chapter has high-lighted many different examples of unethical practices, arguably more frequent in developing countries, which can be very harmful to the outcome of construction pro-jects and ultimately their respective economies (Inuwa et al. 2015).

In the case study of Nigeria, Inuwa et al. (2015) reported that there is a general lack of punishment for unethical practices including corruption and this is one of the major factors that has been responsible for the state of the construction industry. Other prob-lems that have historically played a part in unethical practices include the lack of conti-nuity in government programmes, loss of money through changes in government and political leadership and availability of loopholes in project monitoring. Causes of cor-ruption can be exacerbated by factors such as greed, poverty, politics in the award of building contracts, professional indiscipline and favouritism. Inuwa et al. (2015) articu-lated that in such developing countries the construction industry is more susceptible to ethical problems than that of other industries. This could be explained by the value of the construction industry, and the potential financial benefits from corruption. There are also many different stages in the process from planning a project to completion on site, and each one has many different stakeholders involved who could participate in unethical practices. The tendering process has been long associated with corruption, especially where officials representing clients are empowered to award-building pro-jects to contracts in exchange for bribes.

5.13   Summary

The construction industry provides a perfect environment for ethical dilemmas, with its low-price mentality, fierce competition and paper-thin margins. In addition, different cultures exist within the construction industry, and this may make boundaries between ethical and non-ethical behaviour become blurred at times. Categories of unethical behaviour could include breaches of confidence, conflicts of interest, fraudulent prac-tices, deceit and trickery, presenting unrealistic promises, exaggerating expertise, con-cealing design and construction errors and overcharging. Less obvious examples of unethical behaviour could emanate from the misuse of power within organisations, and this could relate to managers making unfair or unreasonable demands on their subordi-nates, displaying intimidating behaviours or undermining their staff. There are degrees of understanding of what constitutes ethical compliance around regulations and rules, especially where tendering and procurement practices are concerned. Accordingly, one party may consider one practice to be acceptable and ethical, whereas another may find it totally unacceptable and unethical. This demonstrates that what constitutes unethical behaviour in practice is not always black and white and can come in many shades of grey owing to interpretation difficulties. This is particularly the case where gifts and corporate hospitality are concerned, albeit The Bribery Act 2010 does offer some clarity in the area; making it an offence to accept or give or receive in kind inducements to receive favours.

Some have argued that new players in the construction industry could be emerging with a general lack of appreciation of construction ethics. Notwithstanding this premise, causes of corruption can be exacerbated by factors such as greed, poverty, politics in the award of building contracts, professional indiscipline and favouritism. Corruption is particularly a problem in some developing countries where as much as 10% of total global construction expenditure was lost to some form of corruption. Notwithstanding this premise, corruption is not confined to developing countries and financial irregularities, scandals, blacklisting and what could the press have referred to as 'dirty tricks' have brought the UK construction industry into disrepute in recent years. Where corruption has occurred, it can be very damaging for the reputation of those organisations involved, especially where they have been found guilty of breaching ethical standards and committing criminal offences. Many cases of financial impropriety, false reporting and fraud have entered the courtroom sometimes resulting in fines, and in extreme cases imprisonment. The case of construction and professional services company Sweett Group PLC in February 2016 was one such example on a global scale, where they were ordered to pay £2.25m after a bribery conviction. Other non-fine–related consequences for unethical breaches could include damage to reputation (personal and organisational), liberty (maximum 10 years prison sentence) and the inability to participate in tendering for public contracts. It is perhaps not surprising that the reputation of the construction industry has been tarnished over many years, which has culminated in a general lack of public trust and confidence in some countries.

Conflicts of interests have links with corruption and occur where there is a clash of commercial and private interests by individuals who hold a position of authority and trust. It could also be synonymous with an individual who has interests or obligations to two or more parties to a contract. Negligence in the design and construction of buildings represents another type of unethical action. Where contractors or designers have compromised the health, safety and well-being of individuals, including members of the public, through poor design or construction quality this could have catastrophic results with potential loss of life. Some examples of such cases include the partial collapse of Ronan Point tower block in 1968 and more recently the devastating fire spread at Grenfell Tower in 2017.

To reduce corruption across the world and restore trust in politics, it is recommended that global governments reinforce checks and balances around financial auditing, promote separation of powers, tackle preferential treatment, reduce biased towards special interests and prevent excessive money and influence in politics. Finally, other recommendations include managing conflicts of interest, promoting open and meaningful access to decision-making and empowering citizens whilst protecting activists, whistleblowers and journalists.

References

Abdul-Rahman, H., Abd-Karim, S.B., Danuri, M.S.M., Berawi, M.A., and Wen Y.X. (2007). Does professional ethics affect construction quality? Quantity Surveying International Conference. *Kuala Lumpur, Malaysia*. Retrieved from https://www.sciencedirect.com [Accessed November 24th, 2019].

Challender, J. (2019). *The Client Role in Successful Construction Projects*. London: Routledge.

Daily Mail (2020). Grenfell builders shunned safer cladding to save cash. *Daily Mail* 24th October 2020.

Daily Mail (2021a). Building firm's £15bn bonanza since Grenfell. *Daily Mail* 6th January 2021.

Daily Mail (2021b). Deadly secret of Grenfell firm's panel test failures. *Daily Mail* 17th February 2021.

Delbridge, A. and Bernard, J.R.L. (2000). Macquarie Dictionary, Macquarie Point.

Fewings, P. (2009). *Ethics for the Built Environment*. Oxon: Taylor and Francis.

GIACC Global Infrastructure Anti-Corruption Centre (2008). Examples of Corruption in Infrastructure *Examples of corruption in infrastructure*. from www.giaccentre.org [Retrieved November 6th, 2010].

Inuwa, I.I., Usman, N.D., and Dantong, J.S.D. (2015). The effects of unethical professional practice on construction projects performance in Nigeria. *African Journal of Applied Research (AJAR)* 1 (1): 72–88.

Liu, A.M.M., Fellows, R., and Nag, J. (2004). Surveyors' perspectives on ethics in organisational culture. *Engineering, Construction and Architectural Management* 11 (6): 438–449.

Stansbury, C.S. (2008). *Examples of Corruption in Infrastruture*. London: Global Infrastructure Anti-Corruption Centre.

Usman, N.D., Inuwa, I.I., and Iro, A.I. (2012). The influence of unethical professional practices on the management of construction projects in northeastern states of Nigeria (3)2. 124–129.

Vee, C. and Skitmore, M. (2003). Professional ethics in the construction industry. *Engineering,Construction and Architectural Management* 10 (2): 117–127.

Walker, A. (2009). *Project Management in Construction*. Oxford: Blackwell Publishing Ltd.

Weiss, J.W. (2003). *Business Ethics: A Stakeholder and Issues Management Approach*, Third ed. Ohio: Thomson South-Western.

6

Regulation and Governance of Ethical Standards and Expectations

If you are guided by a spirit of transparency, it forces you to operate with a spirit of ethics. Success comes from simplifying complex issues, address problems head on, being truthful and transparent. If you open yourself up to scrutiny, it forces you to a higher standard.

Rodney David

6.1 Introduction

In consideration of the above quotation, this chapter will focus on the importance of organisational transparency, regulation and governance in the pursuit of ethical compliance. This is especially the case following some global high-profile cases of breaches of ethical standards in recent years. However, maintaining such regulation has not always been straight forward across an industry so diverse and encompassing many different professionals. Accordingly, this chapter will explain why traditional self-regulation through professional bodies may not be enough to ensure strict adherence to ethical principles, rules and regulations where such differentiation exists.

Management governance processes and procedures will be discussed alongside organisational financial regulations, and how they are designed to reduce ethical risk associated with breaching established protocols and potentially committing fraud. Delegated authorities and governance structures will be explained alongside the importance of hierarchy committees for various levels of approval and decision making.

Finally, governance, and best practice procurement specifically at the tendering and award stages of contracts will be examined, as this is a potential area where unethical or improper conduct can lead to challenges and legal proceedings. In addition, EU regulations (OJEU) will be highlighted, and the thresholds associated with the value of contracts procured.

6.2 The Problem of Maintaining Standards around Professional Ethics

Over many years the enforcement of standards and adoption of ethical principles has become increasingly important. This has potentially resulted from a reported rise in allegations of unethical behaviours from professionals in construction-related

Professional Ethics in Construction and Engineering, First Edition. Jason Challender.
© 2022 John Wiley & Sons Ltd. Published 2022 by John Wiley & Sons Ltd.

industries including surveying, engineering and project management (Abdul-Rahman et al. 2010).

According, the governance and regulation of ethics should be focused on maintaining standards for all professionals practising within their respective areas of the construction industry. They should, according to Carey and Doherty (1978), encompass expectations from the public which includes responsibility, accountability, competence, honesty and willingness to serve the public. Notwithstanding this assertion, in the past there has been dilemmas around how to regulate and govern such standards and expectations. Traditionally this has relied on self-regulation through professional institutions or governing bodies where behaviour of construction professionals has been controlled and monitored by a strict code of ethics (RICS 2003). These codes of conduct are primarily to ensure that integrity and ethics are maintained at all times will be covered in more detail in Chapter 10. This self-regulation potentially creates potential issues around autonomy of those institutions in that they regulate themselves effectively without any external intervention and thus ensure that their clients are adequately protected. Furthermore, according to Uff (2003) each profession and discipline within the construction industry including architects, surveyors and engineers operate under their own ethical codes in the field of construction law. This creates another problem for maintaining ethical standards and expectations and there are many examples where clients have not benefitted from the actions of professions. In these cases, ethical quality control has been found to have fallen short in terms of projects not being completed on time and budget and to an acceptable level of quality, resulting in low client satisfaction levels. This dilemma possibly presents an argument for a 'multi-disciplinary' approach to ethics for maintaining consistency of design teams. Arguably there is also a need for ethics to be applied in construction law as there is a requirement to regulate the conduct of such professionals to protect their clients and the public (Uff 2003).

6.3   Financial Regulations, Governance Policies and Delegated Authority

Without strong watchdog institutions, impunity becomes the very foundation upon which systems of corruption are built. And if impunity is not demolished, all efforts to bring an end to corruption are in vain.

Rigoberta Menchu, Nobel Prize Laureate

6.3.1   Financial Regulations

The regulations that organisations introduce provide the necessary framework for financial governance. The rules should be operated transparently, consistently and designed to maintain the integrity of companies and their employees (Fewings 2009). There can be severe consequences for breaches and depending on the nature and severity of the breach it could be construed as is misconduct, gross misconduct and/ or a criminal offence. Financial rules and regulations should be regularly reviewed to assess whether they meet the changing environments that organisations work within. Those financial rules and regulations that were introduced in the past and which companies have worked within for many years may simply not be fit for the future, given the changing times and complexity of business dealings. Accordingly, these should be updated as necessary following consultation with internal and external stakeholders. In such circumstances, rules will no doubt be updated and changed, and it is normally the onus of the employees to read and comply with them. Some organisations make it mandatory, when changes are made for staff to undertake an induction to the new regulations possibly via an online module. This is designed to ensure that they understand new rules and regulations and it is not unusual for companies to make employees sign an undertaking to adhere to them. Companies do this to ensure that employees cannot plead ignorance to the changing rules, as a defense for inadvertently breaching the regulations.

6.3.2   Financial Governance Policies

Financial governance policies should set out responsibilities and accountabilities that cover different roles within organisations. According to Fewings (2009) employers normally place responsibilities on senior staff to enforce the financial rules and regulations through agreed procedures and processes under the premise that 'with power comes great responsibility'. These senior individuals should report to the organisation's Head of Finance or Governance on any concerns that may have alongside any possible breaches of procedures and policies.

Financial governance policies and systems should be designed to address risks around ethical compliance. These are normally supported by policies that are developed and employees required to adhere and sign up to. Some examples of such policies are included in the Appendices and include the following:

- Anti-Bribery Policy (Appendix A)
- Counter Fraud and Response Policy (Appendix B)
- Criminal Finances Act Policy (Appendix C)

Arguably there are three main sources of risk for ethical compliance, namely perceived pressure, perceived opportunity and rationalization and these are shown in Figure 6.1. Perceived pressure normally emanates from the desires of employees to conduct business transactions in a timely manner which can sometimes bypass procurement procedures. The pressures and temptations to cut ethical corners and to continue questionable practices can be a constant source of difficulty for companies. An example of this could be where an employee instructs a contractor to carry out works that have not been authorised and budgeted for, owing to a perceived pressure to deal with an urgent repair. Budgets are normally agreed at the beginning of the financial year and are normally monitored and approved by finance departments within organisations and this action to overspend and undertake an unauthorised transaction could be regarded as a flagrant breach of process.

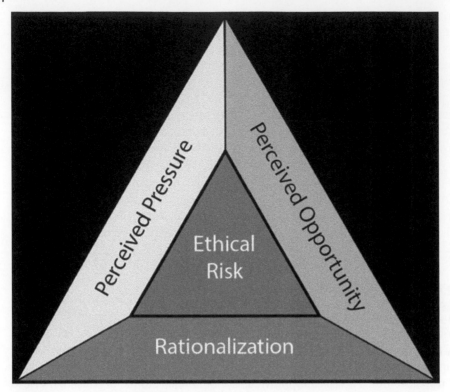

Figure 6.1   The three main areas of ethical risks.

Ethical risk can sometimes culminate from a perceived opportunity that an individual may take to benefit directly from an action which is regarded as unethical. An example could be claiming expenses for a transaction or payment that they have not actually incurred. The third ethical risk namely rationalization is related to ethical justifications; a scale for apprising an individual's reasons for not behaving ethically. An example may be an employee who uses their company credit card to pay for their own groceries, on the basis that they will pay the money back at a later stage.

In a concerted attempt to deter employees for breaching these ethical dilemmas, companies will normally impose rules and regulations related to purchase cards and raising purchase orders. Employers should make it abundantly clear to their staff that purchasing cards are wholly the responsibility of the holder and cards must not be loaned or delegated to others. Furthermore, it should be communicated expressly that these cards must only be used for the purposes they have been issued, and never for personal expenses. It is not uncommon for employees to be disciplined or even dismissed for using such cards for non-business transactions.

With regard to purchasing goods and services for companies, it is normal for thresholds to mandate the procedures that need to be followed depending on the value of those goods and services. An example in the public sector of such thresholds is illustrated in Table 6.1. In addition, companies normally regulate those invoices that require a purchase order to be raised in advance. This safeguards against individuals inadvertently

Table 6.1 Example of procurement requirements for differing values of work orders.

Value of Work Order	Procurement Requirements
Low value between £5,000 and £25,000	3 quotes required
Medium value over £25,000 to £181,302	Tender (via purchasing/procurement department)
High value over £181,302	EU tender through OJEU process

verbally instructing work orders before such work orders are approved through the finance systems of their respective organisations. This is normally regarded as not only an ethical breach but a financial breach of rules and regulations with possible disciplinary consequences.

6.3.3   Financial Delegated Authority

It is normal, certainly for public sector organisations, to have delegated authority limits for different individuals, boards and committees. This is to protect organisations from giving individuals the control over large sums of money and allows the approval of large capital projects to be made by those relevant committees. These delegated authority thresholds are also a mechanism to prevent the large-scale abuse of budgets and expenditure and to deter unethical practices and corruption. Examples within a university context of these delegated authority levels and those individuals and boards who are required to approve those levels of expenditure are show in Tables 6.2 and 6.3.

6.4   Ethical Governance at Tender and Appointment Stages

As previously articulated in earlier sections of the book it is critical for employers to have ethically procedures in place during tendering and appointment stages for those contractors and consultants that they are planning to appoint. Having strict governance

Table 6.2 Example in the university sector of levels of delegated authority for major capital projects.

Role, Board or Committee	<£100 K	<£1 m	<£5 m	>£5 m
Deans of Schools or Director of Professional Service or Pro-Vice-Chancellor	Y	Y	Y	Y
Head of Procurement	Y	Y	Y	Y
Executive Director of Finance		Y	Y	Y
Vice-Chancellor's Executive Team			Y	Y
Finance and Resources Committee				Y
Governing Council				Y

Table 6.3 Example in the university sector of levels of delegated authority for normal recurrent budgeted expenditure.

Role	<£2 K	>£2 K <£20 K	>£20 K <£100 K	>£100 K
Local delegated budget holder	Y			
Manager within a school or professional service		Y	Y	Y
Normal recurrent budgeted expenditure				
Deans of School or Director of Professional Service or Pro-Vice-Chancellor			Y	Y
Head of Procurement			Y	Y
Executive Director of Finance				Y

measures and transparency is therefore critical. There should also be robust pre-determined criteria for evaluation and scoring and each stage of the process should be carefully documented. On the latter point, having records of each stage of the process is very important especially when the awarding organisation is public sector. Challenges from unsuccessful bidders can be brought if they feel they have been unfairly rated and the process may be called into question, or a judicial review requested. By evaluating contractors and consultants on pre-determined criteria, it should allow employers to consider the quality of bids as well as the price. It is still the case in the UK that employers, despite government reports (Egan 1998, 2002; Latham 1994), are still opting for the lowest price without considering other features of bids such as the project team, experience, expertise, specialisms, financial status and suitability for the nature of the project (Challender 2017).

There is an ethical dimension to tendering, when contractors are required to contribute large sums of money and time to formulating bids that they stand little chance of winning. Such costs must be built into their tender prices and accordingly these costs will ultimately have to be paid for by employers. For this reason, wherever possible it is sensible and ethical to limit the number of tendering contractors to an appropriate level for them to invest the time and costs and stand a reasonable chance of success. On most projects this could range from a tender list of four to six bidding contractors. A lower number could mean that clients are not allowing enough competition and the process could be deemed as 'a closed shop' and open to criticism. A higher number of tenders than 6 could be judged as being excessive and it is not uncommon for lists of 10 or more contractors to be invited to tender. This means effectively there is less than a one in 10 chance of success for contractors bidding, which could be classified as inappropriate given the work necessary to prepare the tender submission. Large numbers of tendering contractors could result from 'open tender' processes, rather than restricted tender processes. Examples of portals which enable open tendering are Chest and Intend and these allow unlimited numbers of contractors to submit a tender for a project advertised on their websites. It is not always in the clients' interests to seek contractors' bids in this way as it could involve

Table 6.4 *OJEU* procurement thresholds for 2020/2021.

	Supply, Service and Design Contracts	Works Contracts
Central government	£122,977	£4,733,252
Other contracting authorities	£189,330	£4,733,252
Small lots	£70,778	£884,720

sifting through a large number of submissions and receiving tenders from contractors who are not really suitable for the nature of a particular project.

Arguably the best way to secure the right number of bids and from contractors who have already been through rigorous quality evaluation is to tender through a recognised national procurement framework. Contractors on such frameworks are normally grouped into different categories according to the type a of work they undertake. Examples of these categories could include maintenance, refurbishment or new build and with different sub-categories to represent different bands of project values. In this way clients tendering through the framework can be more confident that contractors are placed in the right category are herein suitable for a particular project. The only ethical downside of frameworks is that there is normally a large cost and commitment of resources required by those contractors applying for inclusion on the framework. Accordingly, there is an argument that this excludes smaller contactors who may not have the resources or be able to afford the costs associated with the process. Examples of frameworks in the UK include the North West Construction Hub (NWCH), Construction Impact Framework (CIF), Yorbuild and Procure North West.

For large public sector projects in Europe there are EU Procurement Directives which must be abided by. The most common directive for public sector procurement of goods, services, works and utilities is known as the *Official Journal of the European Union (OJEU)*. *OJEU* is the online journal that is home to all public sector contracts that are above a certain value which in May 2020 was £4,733,272. See Table 6.4 for *OJEU* Procurement thresholds for 2020/2021.

If a public authority contract is above this value threshold, then EC Procurement Regulations require that it has to be awarded using an EU/*OJEU* compliant route to market. This can involve publishing the opportunity via *OJEU* and therein allows the opportunity to be opened to companies in the UK and European Community. This is designed to ensure free movement of supplies, services and works and there are around 2500 new *OJEU* notices advertised each week. When an *OJEU* notice in the form of an Invitation to Tender (ITT) is released, suppliers can officially apply for the opportunity by using Tenders Electronic Dailey (TED). Each procurement procedure in the *OJEU* has a varying tender process and timelines specify the number of calendar days required between a notice being sent to the *OJEU* for publication and the deadline for submission of responses. A flowchart showing these timelines is included in Figure 6.2.

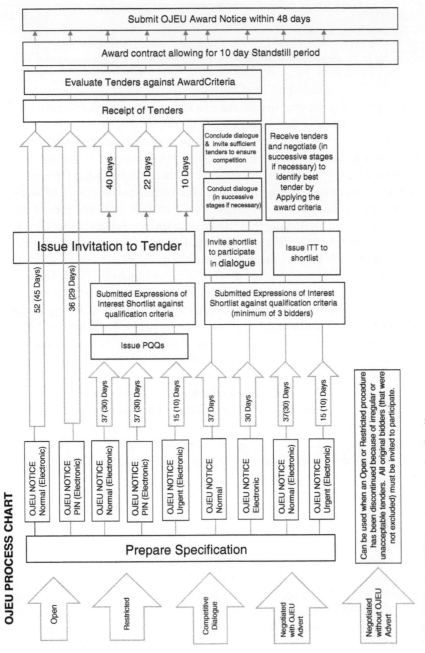

Figure 6.2 An *OJEU* flowchart showing timelines.

6.5 Summary

Regulation of ethics should be focused on maintaining standards for all professionals practising within their respective areas of the construction industry. Such regulation should encompass expectations from the public and include responsibility, accountability, competence, honesty and willingness to serve the public. Notwithstanding this premise, there are dilemmas around how to regulate and govern such standards and expectations. Traditionally this has relied on self-regulation through professional institutions or governing bodies where behaviour of construction professionals has been controlled and monitored by strict code of conduct. However, in some cases, ethical quality control has been found to have fallen short in terms of projects not being completed on time and budget and to an acceptable level of quality, resulting in low client satisfaction levels.

Finance and governance regulations that organisations introduce provide the necessary framework for financial governance. The rules set down within these regulations should be operated transparently, consistently and designed to maintain the integrity of companies and their employees. In addition, they should be updated as necessary following consultation with internal and external stakeholders. Some organisations make it mandatory, when such changes are made for staff to undertake an induction to familiarise them with the new regulations, possibly via an online module. This is designed to ensure that they fully understand the new rules and regulations and it is not unusual for companies to make employees sign an undertaking to adhere to them. Financial governance policies and systems should be designed to address risks around ethical compliance. The pressures and temptations to cut ethical corners and to continue questionable practices can be a constant source of difficulty for companies. In this regard ethical risk could take the form of perceived pressure, perceived risk or rationalisation. In a concerted attempt to deter employees for breaching these ethical dilemmas, companies will normally impose rules and regulations related to purchase cards and raising purchase orders.

With regard to purchasing goods and services for companies, it is normal for thresholds to be created to mandate the procedures that need to be followed according to the value of those goods and services. In addition, companies normally regulate that all invoices require a purchase order to be raised in advance. This safeguards against individuals inadvertently verbally instructing work orders before such work orders are approved through the finance systems of their respective organisations. It is normal, certainly for public sector organisations, to have delegated authority limits for different individuals, boards and committees. These delegated authority thresholds are also a mechanism to prevent the large-scale abuse of budgets and expenditure and to deter unethical practices and corruption.

It is critical for employers to have ethical procedures in place during tendering and appointment stages for those contractors and consultants that they are planning to appoint. Maintaining records of each stage of the tendering process is very important especially when the awarding organisation is public sector, as challenges by unsuccessful bidders may necessitate documentation being presented in the public domain as part of the audit trail. By evaluating contractors and consultants on pre-determined criteria, it should allow employers to consider the quality of bids as well as the price. There is an ethical dimension to tendering, when contractors are required to contribute large sums of money and time to formulating bids that they stand little chance of winning. For this reason, it is both sensible and ethical to limit the number of tendering contractors to an appropriate level for them to invest the time and costs and stand a reasonable chance of success. Utilising recognised national procurement frameworks are regarded in the UK as a reputable and robust instrument for managing the award of contracts. Consultants and contractors on such frameworks are normally grouped into different categories according to the type of a work they undertake. In addition, for large public sector projects in Europe there are EU Procurement Directives which have to be abided by. The most common directive for public sector procurement of goods, services, works and utilities is known as the *Official Journal of the European Union* (*OJEU*).

The next chapter will focus on ethical project controls in construction management as a practical means to comply with some of the governance and regulatory challenges in construction management.

References

Abdul-Rahman, H., Wang, C., and Yap, X.W. (2010). How professional ethics impact construction quality: Perception and evidence in a fast-developing economy. *Scientific Research and Essays* 5 (23): 3742–3749.

Carey, J.L. and Doherty, W.O. (1978). *Ethical Standards of the Accounting Profession.* New York: American Institute of Certified Public Accountants.

Challender, J. (2017). Trust in Collaborative Procurement Strategies. *Management, Procurement and Law Proceedings of the Institution of Civil Engineers* 170 (3): 115–124.

Egan, J. (1998). *Rethinking Construction. The Report of the Construction Task Force.* London: DETR.TSO. 18–20.

Egan, J. (2002). *Accelerating Change. Rethinking Construction.* London: Strategic Forum for Construction.

Fewings, P. (2009). *Ethics for the Built Environment.* Oxon: Taylor and Francis.

RICS (2003). *COBRA 2003, Proceedings of the RICS Foundation Construction and Building Research Conference 1st to 2nd September 2003.* University of Wolverhampton: RICS.

Latham, M. (1994). *Constructing the Team.* London: The Stationery Office.

Uff, J.P. (2003). Duties at the legal fringe: Ethics in construction law. *Society of Construction Law*, October 2010. Retrieved from www.scl.org.uk (Accessed 28th May 2018).

7

Ethical Project Controls in Construction Management

It takes less time to do things right than to explain why you did it wrong.

Henry Wadsworth Longfellow

7.1 Introduction to the Chapter

This chapter predominantly is intended to make construction professionals aware of some of the governance requirements on their projects and to present some practical control mechanisms for regulatory adherence. It will explain some of the processes around project approval and the importance of project/programme boards in this regard. Various examples of templates and processes for various stages of the approval process will be given to assist construction professionals understand the level of information required. This is intended as a practical guide or toolkit for encouraging best practice.

7.2 Project Controls

For construction professionals, it is imperative that they understand their roles, responsibilities and authorisation levels when leading their project teams. This will ensure they do not make decisions which have a financial or operational effect on projects which should have been escalated to a higher authority. Normally authorisation levels are established for a given project, or they mirror financial regulations for the client organisation. For construction professionals, having robust decision-making processes whilst ensuring governance procedures are maintained is fundamental; avoiding unnecessary delays whilst not infringing financial regulatory requirements.

7.3 The Importance of Project/Programme Boards

The main purpose of these boards is to provide governance and transparency on projects, and avoiding the responsibility and accountability for decision making falling to one individual. Having project and programme boards should facilitate a cross-disciplinary approach to manging the governance of projects, which is especially

Professional Ethics in Construction and Engineering, First Edition. Jason Challender.
© 2022 John Wiley & Sons Ltd. Published 2022 by John Wiley & Sons Ltd.

important when large capital investment is involved. Furthermore, it also allows for additional resources to be committed to and creates joint accountability for decision making and project progress.

It is normal for project boards to approve various stages of projects and manage any major variations and deviations. Accordingly, decisions can be taken by those individuals who make up the project boards depending on the financial impact. In some cases, if the value of a decision-making process exceeds the authorisation level of the project board then the decision making needs to be escalated to a higher level. Under normal situations, this higher level could be the client executive team. Notwithstanding this, some public sector bodies, which can include universities or local authorities, sometimes have a hierarchy of boards that need to be reported to for seeking project or programme approval.

One example, in the context of the university sector, is illustrated in Figure 7.1 which shows a Project Board reporting to Programme Board, which in turn reports to a Finance and Resources Committee and ultimately a Governing Council/Board of Governors.

Limits of authority for each board will be clearly set down in governance and financial regulations and examples of these were referred to in the last chapter. Construction professionals need to be mindful when their projects require a higher tier of approval within their respective organisations, that they allow sufficient time in their project programme for the approval process. The reason for this, is that normally higher-level boards will only consider those capital projects that have been approved by lower boards, and for which robust business cases and reports recommending reasons for approval are clearly set down. Where some boards are not convened on a frequent basis, e.g., every quarter, project planning around critical milestones where approval need to be received are crucial for avoiding programming delays.

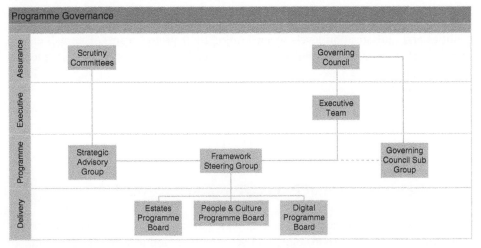

Figure 7.1 Chart illustrating an example of a client approval process across a hierarchy of different project/programme boards.

7.4 Gateway Processes for Project Approval and Business Cases

In considering the approval processes for projects through the individual boards, it is commonplace for robust and financially rigorous business cases to be prepared to support business ventures. In some organisations, mostly public sector, there may be 'gateway processes' which require staged approval depending on the stage and development of projects. An example, taken from a UK university governance process could be as follows:

Obtain Gateway 1 approval

Normally this would be at the RIBA Stage 1, once an initial feasibility study has been prepared and construction professionals are seeking 'in principle' approval to proceed to next design stages, based on feasibility costs, programme and outputs. At this early stage, predictability in terms of design intent, costs and programme may be indicative, and subject to further design development. This is normally supported by an outline business case setting out the various options and the benefits and possibly implications of the preferred option that the Gateway seeks approval for. An example of a template for the Gateway 1 project proposal is contained in Appendix D.

Obtain Gateway 2 approval

Normally this would be at the end of RIBA Stage 2, once initial feasibility studies have been developed into more detailed plans and designs. It would normally be expected that 'composite cost' plans have been prepared at this stage which have estimated costs on a square metre rate, to give approximately 60% cost certainty. It is possible at this stage that more developed 'elemental costs' have been prepared which are based on more detailed cost breakdowns (by elements) and give approximately 70% cost certainty. The extent to which the costs have to be developed for a Gateway 2, really depends on the governance and financial policies set down by a specific organisation. At this stage, construction professionals are seeking approval to proceed to full detailed design stages, which normally implies a commitment to approve the remainder of the pre-construction professional design team fees. This is normally supported by a more developed business case.

Obtain Gateway 3 approval

Normally this would be at the end of RIBA Stage 4, once the detailed design has been completed and tender prices have been obtained. It would normally be expected that tender costs are fixed and based on detailed cost and tender documents for to give 100% cost certainty. At this stage, construction professionals are seeking approval to proceed to the construction phase and as such, a commitment

---

(Continued)

(Continued)

to approve the whole project costs. Normally the 'whole project costs' include all associated costs including vat, professional fees, decant costs, contingencies, equipment and fixtures and fittings. This Gateway 3 as the final gateway must provide a full business case, setting out a developed and robust business case with financial rigour and strong arguments for the development to proceed, having considered all other available options. An example of a template for the Gateway 3 business case process is contained in Appendix E.

When considering the above Gateways 1, 2 and 3, the flowchart in Figure 7.2 could be helpful in illustrating how each of the gateways are integrated into an overall process for approving projects, at the various board levels and design stages.

7.5 Summary

This chapter has articulated the importance of construction professionals having robust financial processes in place. Accordingly, it is imperative that they understand their roles, responsibilities and authorisation levels when leading their project team. Furthermore, it is important for construction professionals to have robust decision-making processes in place within their respective organisations whilst ensuring governance procedures are maintained.

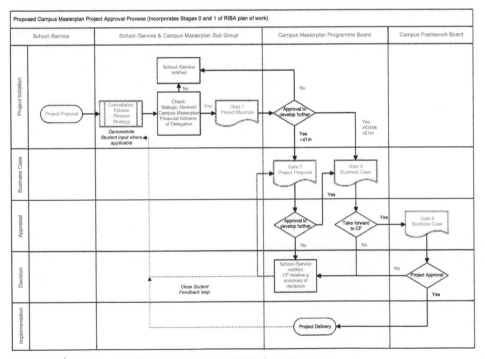

Figure 7.2 Flowchart to illustrate Gateway approval process.

Project and programme boards are normally the vehicle by which the decision-making process is governed. Their main purpose is to provide governance and transparency to projects, and avoiding the responsibility and accountability for decision making falling to one individual. With this in mind, where the value of a decision-making process exceeds the authorisation level of the project board then the decision making needs to be referred to a higher level. Accordingly, construction professionals need to be mindful when their projects require a higher tier of approval within their respective organisations, that they allow sufficient time in their project programme for the approval process.

In considering the approval processes for projects through the individual boards, it is commonplace for robust and financially rigorous business cases to be prepared to support the business venture. In some organisations, mostly public sector, there are gateway processes for project approval and business cases. The gateway processes will highlight different levels of approval for projects depending on which design stage they are at. Practical examples and templates of gateway applications, supported by business cases have been presented in this chapter, to assist readers in this regard.

8

Developing an Ethics Toolkit, as a Practical Guide for Managing Projects

We are trying to move away from looking at ethics as scandal to ethics as strategy. We are lifting the hood and seeing the concrete strategies companies are using as this might help to see what other companies can do.

Peter Singer

8.1   Introduction to the Chapter

This chapter is designed to assist construction professionals in the management of all stages of projects and particularly around legal and ethical compliance. With reference to the above quotation in seeing ethics management as an organisational strategy, proformas and checklists have been specially developed for this chapter to improve the robustness of processes and for compiling important data on projects. These have been devised to ensure that critical aspects of projects, including statutory compliance and quality controls are not compromised during the life of projects. In this way, it is intended as a practical guide to managing the processes and procedures linked to each aspect.

Following Driscoll's (three elements of reflection) Model (Driscoll 2000), the author's reflective practice also led to reflective enquiries around construction professionals and specifically leading projects from an ethical and compliance standpoint. This triggered several questions which the author sought to unravel. These included:

- Where is the evidence that justifies the idea that theory and practice need to come together in ethics and compliance awareness when leading and managing construction projects?
- Why is there a gap in provision around awareness of ethics knowledge, training and guidance for construction professionals?
- What improvements could be made to bridge the gap?

In consideration of the above questions, and from interviews with construction professionals for this book, it has become apparent that some, especially those who lacked previous construction experience, feel ill-equipped with the basic skills specifically around ethics and compliance required to successfully procure projects. Furthermore,

Professional Ethics in Construction and Engineering, First Edition. Jason Challender.
© 2022 John Wiley & Sons Ltd. Published 2022 by John Wiley & Sons Ltd.

these individuals articulated that they relied heavily on their design consultants to guide them through the necessary steps and processes. Their lack of expertise and training in professional practices around construction management led the authors to consider a 'toolkit' which could capture the various stages of projects and offer practical guidance at each stage. The intention of this guidance document would be predominantly to bridge gaps in knowledge and ensure that critical stages of projects are given sufficient consideration by professionals. The overarching benefit would be to reduce the risk that construction professionals could inadvertently breach ethical standards around due diligence and process.

8.2   Planning and Devising the Toolkit

As far as addressing the gap in knowledge, available guidance documents and general provision around raising awareness and competencies of construction professionals, it was considered useful to develop an innovative and meaningful teaching and learning resource. The primary aim was to enhance the existing skills and awareness of professionals as a knowledge base. It would seek to benefit professionals when procuring projects and enable construction professionals to improve their leadership and management skills to promote successful outcomes.

- Create a teaching and learning 'step by step' practical guide that will assist professionals through the construction process and allow them to lead their project teams in a professional and proactive way.
- Familiarise construction professionals with some of the necessary processes and governance requirements for projects to better prepare them for procuring projects.
- Close current gap between those existing education/training skills gained by construction professionals and those required when procuring projects.

A project plan proposal form was prepared by the author of this book which mapped out the aims, objectives of the practical guide, to assist in its creation. It was essentially designed to focus on the development of a simple and concise practical guide through the various procurement stages. This was intended as a guide and template to assist construction professionals in undertaking building projects and leading them on a professional capacity. Its main purpose was intended to steer construction practitioners through a logical sequence as far as the various stages and the methodology of the construction management processes are concerned from start to finish. In this way it is hoped that risks associated with 'shortcuts' in processes, procedures and documentation management will be reduced and an ethical and robust project management platform established.

The basis of the practical guide or 'toolkit' in this chapter is to provide a reference tool to successfully procure projects and designed to give all the necessary information that construction professionals require to enable them to comply with good practices. It also would provide construction professionals with a basic familiarity and understanding on some of the procurement 'checks and balances' that should be considered. This will then hopefully benefit them in their projects and better prepare and equip them for some of the challenges faced.

8.3   Feedback and Evaluation of the Toolkit from the Perspectives of Construction Professionals

A consultation through interviews with construction professionals was carried out to review the 'toolkit' and obtain their completed feedback on how useful they found the practical guide.

Positive feedback from all participants included the following:

- The practical guidance is clear and unambiguous.
- The various checklists, templates and proformas contained within the toolkit allow building information to be easily compiled throughout the life of projects.
- The toolkit promotes a logical and methodical step-by-step approach to carrying out building projects and helps to ensure that ethical issues around compliance and documentation management are not breached.
- As a practical guide it should ensure that each stage of the construction procurement process is undertaken in a systematic way with no component parts left out. This is intended to avoid risky short cuts which could have very damaging and costly implications.

Notwithstanding the above feedback, the advantage from an educational perspective is that construction professionals reported that they feel they are now more prepared and able to progress projects more confidently, using the information compiled within this practical guide. The guide is intended to steer them through the construction management processes, stage by stage, and allow them to comprehend and manage the various aspects of projects. Furthermore, it is felt that the guide could be conducive as a teaching and learning tool for industry. In its existing simplistic form, it could potentially save a lot of time for professionals in collating information more swiftly.

The following subchapters articulate the various stages of projects and aids construction professionals through templates, checklists and proformas as part of a practical guide or 'toolkit'.

8.4   Ensuring and Monitoring Performance Throughout the Life of Projects: General Project Directory and Checklist

One of the important considerations for construction professionals is to keep data on projects on one system or folder whereby information on projects can be recovered and updated easily. There should ideally be an overarching project checklist which should cover all stages of projects from inception to completion. This 'master checklist' is intended to capture other checklists at various stages of projects and therein be a useful management tool as projects progress. An example of a general project checklist is included in Figure 8.1.

PRE-CONSTRUCTION PROGRESS REPORT

Report Details

Project Title:

Report No. | Report Date: | Time Elapsed:

Overview and further items for discussion at the next progress meeting.

Programme — Status:

Accepted Programme Revision:		Date of the Last Issued Programme:	
Next Programme Revision:		Date to be Issued by:	

Pre-Construction		Construction		Status Notes & Actions
Start Date Pre-Con		Start Date Construction		
End date Pre-Con		End date Construction		
Planned Pre-Con completion		Planned Construction completion		
Variance:		Variance:	weeks	
PSC Period (calendar wks.)		Main Works Period (calendar wks) (Decant excluded)	weeks	

Budget — Status:

Pre-Construction (contract stage 1)		Construction (contract stage 2)		Status Notes & Actions
Original Pre-Con £ (end of RIBA 4)		Original Construction £ (RIBA 4-7)		
Forecast Pre-Con £ (end of RIBA 4)		Forecast Construction £ (RIBA 4-7)		
Variance:		Variance:		
Total MS (Contract Stages 1 + 2)		**Total Client (Contract Stages 1 + 2 + Client £)**		
Original total £ (RIBA 2-7)	£ -	Client affordability Gross £		
Forecast total £ (RIBA 2-7)	£ -	Forecast Gross £		
Nett Variance:	£ -	Previous agreed Gross £		
Cost plan ref		Gross variance to Client affordability:	£ -	Current Risk Allowance:

Design — Status:

				Status Notes & Actions
RIBA Stage		Client target gross area m2		
End of stage date		Forecast gross area m2		
Weeks left in this stage		Variance to gross area m2		
Feasibility Nett/Gross %		Proposed Nett/Gross ratio %		

Contract & Orders — Status:

				Status Notes & Actions
Pre-Construction Contact Status:		Contract Type:		
Main Contact Type:				
Draft Contract receipt date:		Anticipated Contract execution date:		

Early Warnings — Status:

				Status Notes & Actions
Number of EW Issued:		Number of EW open:		
Number of CE Issued:		Number of CE open:		

Description of Early Warning notices issued in the reporting period:

Figure 8.1 General project checklist.

8.5 The Documentation that Construction Professionals Need to Consider at Pre-Construction Stages

There are many different documents that professionals should source from their construction partners prior to works commencing on site. These are largely related to health and, compliance and competency.

8.5.1   Contractor Pre-Questionnaires for Competency and Compliance

Questionnaires are sometimes commonplace. These are designed and implemented by construction professionals to assess contractors' suitability prior to them being included on tender lists. Construction professionals normally adopt schemes such as the CSCS scheme and their questionnaires relate to pre-qualifying criteria for such affiliated accreditation. Contractors not part of these schemes must apply in writing and complete the relevant health and safety competency questionnaire. These are normally approved by the construction professionals prior to commencing the tender processes.

On receipt of the contractor pre-qualification information, this should be forwarded to the relevant client officer who will add the details to the 'Contractor Compliance List' and advise the construction client if the tender process can now progress. In some circumstances, it may be necessary for contractors to become members of a recognised scheme, due to the specialist nature of the works to be completed. These could include schemes such as the Considerate Contractors accreditation, linked to community liaison and ensuring neighbouring parties are not unduly affected by construction projects. An example of a pre-qualification competency questionnaire is contained in Appendix F.

8.5.2   Monitoring Checklist Required for Document Control and Processes

Following the construction professionals' tender processes leading to contractor selection, it is essential for construction professionals to ensure that their contracting partners are checked to have the right compliance measures in place, prior to commencing work. Figure 8.1 has been developed as a 'Monitoring Checklist' for construction professionals or their project managers to ensure that all documentation is in place in this regard.

The checklist will serve construction professionals to verify that their contractors have compliance measures in place and are fully inducted before being instructed to commence works. This is designed to ensure that the following provisions have been carried out or in place and therein meet the ethical and legal compliance requirements in Figure 8.2.

8.5.3   Permits to Work and Making Contractors Aware of Known Hazards

Permits must be signed off and returned to appointed officers, nominated by construction professionals, prior to work commencing and at completion. Typically, tasks and activities which may need to be authorised by a Permit to Work will be those that involve hazards, and these are included in Figure 8.3.

8.6   Managing Documentation and Processes Following Appointment of Contractors

8.6.1   Project Execution Plan

A project execution plan (PEP) is the governing document that establishes the means to execute, monitor and control projects. The plan serves as the main communication vehicle to ensure that everyone is aware and knowledgeable of project objectives and how they will be accomplished. A project execution plan (PEP) will typically contain information as set down in Appendix G.

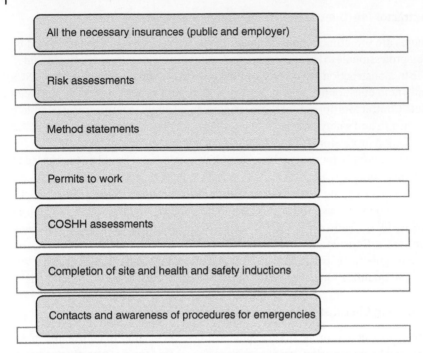

Figure 8.2 Ethical and legal compliance requirements checklist for contractors, prior to commencing on site.

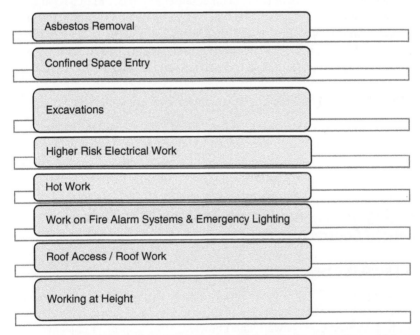

Figure 8.3 Typically, tasks and activities involve hazards for contractors which may need to be authorised by a Permit to Work.

8.6.2   Contractors' Health and Safety Handbooks and Codes of Conduct

In addition to above, it is good ethical practice for construction professionals to compile their own health and safety contractors' handbooks, which is normally tailored to their particular organisations and estates. These should incorporate important information on what contractors need to know about construction professionals' sites and buildings and safety precaution that must be strictly adhered to. An example of a contractor's health and safety handbook is included in Appendix H. It is also common in these health and safety handbooks for contractors to have to sign up to an employer's code of conduct and site rules. These should ensure that contractors comply with the rules, regulations and procedures when entering the professionals' buildings and estates. Any shortcomings, in this regard, should be rectified immediately and the overall health and safety performance of a contractor reviewed regularly.

Construction professionals should ideally appoint an officer within their respective organisations to carry out random site inspections to assess compliance with control measures employed by contractors. The frequency of the checks should be appropriate for the size, complexity, location and nature of projects. Clearly those projects that are potentially more complex and have the risk of causing more disruption to professionals' businesses and staff should be managed more closely through such checks and inspections. This is particularly the case when contractors are working in 'live' environments and business continuity represents a major issue for professionals. Professionals have a duty of care to ensure that their buildings and estates are clean and safe for employees, visitors and members to enter into. Any accidents on their respective sites could lead to legal action being taken against client organisations for negligence and duty of care breaches.

8.7   The Documentation and Processes that Construction Professionals Need to Consider in the Construction Phases of Projects

8.7.1   The Importance of Monthly Project Reporting for Construction Professionals

Construction professionals frequently are responsible for more than one projects which may be running concurrently at any time. These projects may be part of a programme of projects, and in some cases where there is considerable capital investment, a large transformational 'masterplan'. This could be particularly relevant to public sector bodies or large private sector professionals embarking on regeneration schemes. This large-scale development normally places undue challenges on construction professionals in being able to have a full high-level perspective on a programme of projects. For this reason, it is important for construction professionals to provide with regular, preferably monthly, high-level reports on progress of projects from their project teams. This should include an exceptive summary, progress against milestones, matters arising since the last report and any financial considerations. Furthermore, it should ideally incorporate a cost plan, reporting actual expenditure against planned, together with a 'dashboard' that could indicate project particulars including risks. An example of a cost plan (blank) is included in Table 8.1 and a progress dashboard template (blank) in Figure 8.4.

Table 8.1 Example of a cost plan.

Cost Report Nr 11

SECTION 2: FINANCIAL SUMMARY & COMMENTARY

2.1 CONSTRUCTION SUMMARY

		This Report (11)	Previous Report (10)	Change + / (-)
1	**Contract Sum**			
2	Adjustment of PC and Provisional Sums (Appendix A)			
3	Adjustment of Day Works Allowance (Appendix B)			
4	Instructed Variations (Appendix C)			
5	Anticipated Variations (Appendix D)			
6	Contractual Claims (Appendix E)			
7	Project Risks (Appendix F)			
8	**Estimated Final Contract Value (Excl. VAT)**			
9	**Anticipated Over / (Under) Spend on Contract Sum [8-1]**			
10	**Client Directly Incurred Costs (Appendix G)**			
11	**Project Costs (Inc Fees & Excl VAT)**			
12	**Project Costs (Inc Fees & Inc VAT)**			
13	**Project Budget**			
14	**Variance Against Budget**			

2.2 COMMENTARY (Change in Period)

(i) This cost report includes the following movement in the reporting period:-

	This Report (11)	Last Report (10)	Change in Period
Instructed Variations (Appendix C)			
Anticipated Variations (Appendix D)			

Basis of Cost Report

(ii)

(iii)

(iv)

(v)

Expenditure

(vi)

Current Issues

(vii)

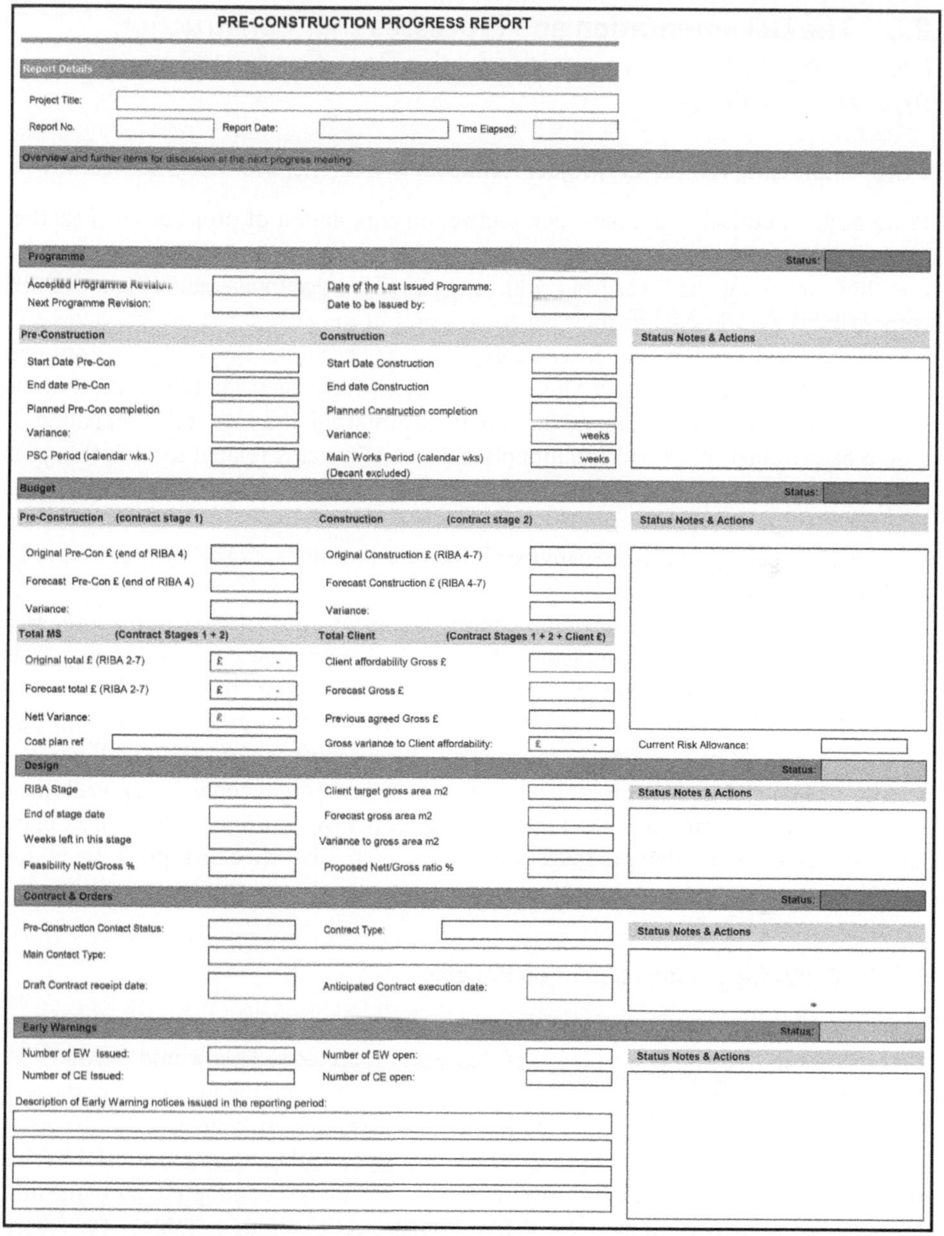

PRE-CONSTRUCTION PROGRESS REPORT

Report Details

Project Title:

Report No. | Report Date: | Time Elapsed:

Overview and further items for discussion at the next progress meeting

Programme Status:

Accepted Programme Revision: | Date of the Last Issued Programme:

Next Programme Revision: | Date to be issued by:

Pre-Construction		Construction		Status Notes & Actions
Start Date Pre-Con		Start Date Construction		
End date Pre-Con		End date Construction		
Planned Pre-Con completion		Planned Construction completion		
Variance:		Variance:	weeks	
PSC Period (calendar wks.)		Main Works Period (calendar wks) (Decant excluded)	weeks	

Budget Status:

Pre-Construction (contract stage 1)		Construction (contract stage 2)		Status Notes & Actions
Original Pre-Con £ (end of RIBA 4)		Original Construction £ (RIBA 4-7)		
Forecast Pre-Con £ (end of RIBA 4)		Forecast Construction £ (RIBA 4-7)		
Variance:		Variance:		
Total MS (Contract Stages 1 + 2)		**Total Client (Contract Stages 1 + 2 + Client £)**		
Original total £ (RIBA 2-7)	£ -	Client affordability Gross £		
Forecast total £ (RIBA 2-7)	£ -	Forecast Gross £		
Nett Variance:	£ -	Previous agreed Gross £		
Cost plan ref		Gross variance to Client affordability:	£ -	Current Risk Allowance:

Design Status:

				Status Notes & Actions
RIBA Stage		Client target gross area m2		
End of stage date		Forecast gross area m2		
Weeks left in this stage		Variance to gross area m2		
Feasibility Nett/Gross %		Proposed Nett/Gross ratio %		

Contract & Orders Status:

				Status Notes & Actions
Pre-Construction Contact Status:		Contract Type:		
Main Contact Type:				
Draft Contract receipt date:		Anticipated Contract execution date:		

Early Warnings Status:

				Status Notes & Actions
Number of EW Issued:		Number of EW open:		
Number of CE Issued:		Number of CE open:		

Description of Early Warning notices issued in the reporting period:

Figure 8.4 Example of a project dashboard template.

The monthly reporting should capture the current risk on projects, identified as part of the project execution plan (PEP) and report against them. Where there are risks, mitigation to control and reduce those identified risk should be documented, and these measures are performing. There should also be reporting on quality management in the build processes and a general overview of whether the building is performing to key performance quality standards and therein fit for purspose.

8.8 The Documentation and Processes that Construction Professionals Need to Consider in the Post-Construction Phases of Projects

8.8.1 Managing Handover: Project Handover Checklists and Test Certification

From both an ethical and a legal perspective, on completion of projects, construction professionals should have robust procedures to ensure that the building or accommodation they are taking handover from contractors is safe to occupy. Arguably some of this responsibility clearly falls to their consultant project managers, but professionals still have due diligence in terms of guaranteeing a safe and compliant working environment for their staff. In practice, many construction projects have been completed without the full array of 'checks and balances' in place to ensure legal and safe staff occupation. It would be extremely easy and commonplace for certain items related to such things as test certification and mechanical and electrical commissioning to be missed at project handover stage. For this reason, the project handover/test certification checklist in Table 8.2 has been prepared to capture all the relevant items, normally required to be signed off prior to completion.

In addition to the test certification checklist, it is also good practice to complete a separate fire safety project completion checklist for health, safety and well-being. This covers such things as fire extinguishers installation, reinstatement of fire alarm detectors and call points, correct signage being in place showing new building layouts and ensuring clear/unimpeded fire escape routes. Table 8.3 has been prepared to illustrate a health, safety and well-being checklist. Construction professionals' responsibilities should include ensuring that the most appropriate persons, normally the project manager or health and safety manager, has signed off this checklist prior to taking handover.

8.8.2 Capturing Lessons Learnt on Projects

Construction professionals should ensure that there is a post-contract period of reflection to review what went well and not that well on projects. This would normally be conducted through workshops across the whole project team, including the end users. This reflection is designed to understand the lessons learnt which can be carried over to future projects, and therein improve future practices. In theory, without establishing those lessons learnt and what improvement measures should be adopted, the same mistakes or bad practices could continue from one project to another. Table 8.4 has been prepared to illustrate a pro-forma that could be used as an agenda throughout lessons learnt workshops.

The lessons learnt pro-forma is split into three parts relating to the different stages of projects, namely pre-construction, construction phase and post-construction. Its primary purpose is intended to highlight areas for improvement and suggest improvement or mitigation measures to avoid the same issues arising in the future. To achieve

Table 8.2 Project handover/test certification checklist.

	DESCRIPTION	RECIEVED	N/A
1.00	**ELECTRICAL CERTIFICATES**		
• Electrical Test Certificates		☐	☐
• Fire Alarm Test Certificates		☐	☐
• Emergency Lighting Test Certificates		☐	☐
• Lift Commissioning Test Certificates & Manual		☐	☐
2.00	**MECHANICAL CERTIFICATES**		
• Gas Suppression Installations		☐	☐
• Natural Gas Installation Certificates / Schematic Drawings		☐	☐
• HV Installations and Sub-stations Manual		☐	☐
• Chlorination Certificate(s)		☐	☐
• Air Cooling and Refrigeration Plant Commissioning*		☐	☐
• Air Handling Units, Boilers and Plant Room Equipment		☐	☐
• Fume Cupboards and Local Extract Ventilation		☐	☐
• Building Management System.		☐	☐
3.00	**ITS CERTIFICATES**		
• Data Installation Test Certificates		☐	☐
• AV Installation Commissioning		☐	☐
4.00	**OTHER**		
• Lightening Protection Commissioning Certificates		☐	☐
• Fire Door and Fire Stopping Certificates		☐	☐
• Access Equipment Test Certificates		☐	☐
• Energy Certificates – EPC and DEC		☐	☐

full value from the lessons learnt exercise, a series of improvement measures should be formulated and an implementation strategy by which such measures can be integrated into projects. This normally involves reforming existing project management processes and procedures to capture the improvement measures at the relevant project stages.

Table 8.3 Fire safety project completion checklist.

Fire Safety Project Completion Checklist.

Project Officer

Project Title

Project Code

Completion Date

Have all fire alarm detectors identified on the FRIP been reinstated?			Y/N
Have Fire extinguishers been replaced / installed as per the Fire Strategy document?			Y/N
Has the following fire signage been replaced using photo-luminescent signage;	Manual call point		Y/N
	Fire extinguishers		Y/N
	Do not use lift		Y/N
	Door override buttons (green break glasses)		Y/N
Have fire action notices been replaced?			Y/N
Have fire escape routes been maintained and signage provided (these must be illuminated in licensed premises)?			Y/N
Have refuge points been resigned and communications maintained?			Y/N
Have fire doors been maintained to be opened with one action and both leafs available (bolts must not be fitted to these)?			Y/N
Has fire stopping been installed and checked on any penetrations to walls?			Y/N
Have building plans been updated showing new layouts, etc.?			Y/N

Table 8.4 Lessons learnt proforma.

Stage 1, Prior to commencement of the works:	YES	NO	NOT APPLICABLE OR UNSURE
1 Did we make effective initial contact with you?			
Comments:			
2 Did we communicate with you effectively?			
Comments:			

Stage 1, Prior to commencement of the works:	YES	NO	NOT APPLICABLE OR UNSURE
3 Were we supportive of the aims of your department?			
Comments:			
4 Did we explain the project budget in appropriate detail?			
Comments:			
5 Did we fully understand your department's functions & critical requirements?			
Comments:			

Stage 2, During construction/refurbishment	YES	NO	NOT APPLICABLE OR UNSURE
6 Was the planning and phasing of work effective?			
Comments:			
7 Were you involved in decision making where appropriate?			
Comments:			

Stage 2, During construction/refurbishment	YES	NO	NOT APPLICABLE OR UNSURE
8 Did we respond to your requests/queries in a timely manner?			
Comments:			
9 Did we give you appropriate early alerts to potential problems?			
Comments:			
10 Was any disruption of your functions/services anticipated?			
Comments:			

Stage 3 - Completed construction/refurbishment	YES	NO	NOT APPLICABLE OR UNSURE
11 Did the completed work comply with the agreed brief?			
Comments:			
12 Did the completed work meet your accommodation needs?			
Comments:			

Stage 3 - Completed construction/refurbishment	YES	NO	NOT APPLICABLE OR UNSURE
13 Did the completed work provide an environment that satisfies its users?			
Comments:			
14 Did we deliver a completed project that met your expectations taking into account budgetary restraints?			
Comments:			

8.9 Summary and Usefulness of the Toolkit

In general terms the information included in this chapter could become a useful, simple and innovative practical guide for construction professionals, especially those with little or no previous experience in managing projects. It could offer them a useful management tool, as relative beginners to the task of carrying out complex building projects, and hopefully narrows the gap between their existing knowledge base and those skills required for successful construction management. The various proforma and templates, tailored to what is believed construction professionals require, should form the impetus for improved practices as checklists to ensure critical information and processes for ethical and legal compliance are strictly adhered to.

The opportunities for the guidance contained in this chapter could support future development for professional practice and in education context a useful teaching and learning resource or handbook. This is it presents a more effective and efficient means of collating building information and a step-by-step route map through complex procurement stages and processes. It has the potential to become integrated into BIM systems which client organisations may wish to implement. This technology could be networked and downloaded on to a software system that could effectively prepare a report automatically. This would make the whole process so much more efficient and greatly reduce the resources and time currently expended by construction professionals' staff in preparing various reports. This is particularly relevant for client companies seeking alternative ways to reduce costs and become more competitive accordingly.

Reference

Driscoll, M.P. (2000). *Learning for Instruction*, 2e. Florida State University.

9

Ethical Selection and Appointment Processes for the Construction Industry

A lack of transparency results in distrust and a deep sense of insecurity.

Dalai Lama

9.1 Introduction to the Chapter

This chapter will highlight the importance for construction professionals of the ethical processes around selection and appointment of their contracting partners. With reference to the quotation above, if processes and procedure leading to contract award do not have the right level of transparency and sense of impartiality, then this can have very damaging repercussions. Having the right main contractors on board is arguably one of the most important aspects for construction professionals. Accordingly, this chapter will articulate some of the issues that construction professionals need to be aware of when commencing their selection processes. This is intended as a means of bringing rigour and transparency to existing processes and ensuring that a staged approach to the appointment of project teams is not compromised. Furthermore, it will examine the perceived problems that construction professionals face in selecting the right contracting partners, especially in the case of two-stage tender procurement. It also provides a historical perspective on this overarching ethical dilemma for construction professionals, with particularly reference to government reports.

This chapter is focused predominantly on a research study by Sean Taylor (2017), which established what information and ethical methods are available to construction professionals to help them to identify and select suitable contracting partners. This was aimed at promoting the choice of the right partners for a particular project and encouraging long-term collaborative relationships for achieving successful project outcomes. Previous opinions had suggested that construction professionals may not be employing the full use of robust pre-qualification methods in their appointment processes. This could possibly be hindering their chances of selecting the most suitable partners which could ultimately be adversely affecting project outcomes. In this regard, the study provided important findings to support the following areas of research:

Professional Ethics in Construction and Engineering, First Edition. Jason Challender.
© 2022 John Wiley & Sons Ltd. Published 2022 by John Wiley & Sons Ltd.

- Determination of whether the quality of constriction clients' prequalification processes influence project success.
- Determination and analysis of whether the current pre-qualification practices of construction professionals are sufficiently robust to identify and select the right contracting partners.
- Identification of the extent to which construction professionals are using objective assessments of key performance indicators (KPIs) to shortlist contractors for tenders.

9.2 The Importance of the Contractor Selection Process

Critchlow (1998) stressed the importance of the selection processes when using collaborative procurement strategies, in choosing the most appropriate contractors to realise benefits of partnering through pro-activity, building team spirit, employing lateral thinking and exploring alternatives. Gulf Construction (2008) reinforced this view and advocated that the selection processes should carefully evaluate contractors' experience, skills, resources and expertise rather than simply appointing contractors on lowest tender price. Furthermore, Governments' Procurement Group advocated that selection criteria and competition should be on the basis of life cycle value for money rather than simply reliant on initial capital costs (Procurement Group 1999). This view was supported by Kwayne (1991), who advocated that owing to contractors' early involvement with project teams, selection should be firmly based on value rather than lowest tender. Egan also underlined this opinion and stated that:

> *Too many clients are indiscriminating and still equate price with cost, selecting designers and contractors exclusively on the basis of the tendered price.*
>
> Egan (1998)

Selection criteria was also referred to in the Code of Procedure for Two Stage Tendering which recommended, at the short-listing stage, that contractors should be evaluated upon financial stability, competence, general expertise and experience, approach to quality assurance, capacity and the company's technical and management structure. In addition, the UK Governments' Procurement Group suggested that incentives should be incorporated into building contracts to promote the extent and degree of collaboration in yielding further benefits and improvements beyond those included for within the agreement itself (Procurement Group 1999).

Farmer (2016), highlighted that there was still a 'collaboration problem' in the UK and its adversarial approach remains a salient issue, which makes the industry resistant to change. Despite a succession of government and academic reports (Latham 1994; Egan 1998, Egan, 2002), anecdotal evidence would appear to indicate that whilst majority of private clients have accepted the need for, and the potential advantages of partnering and early contractor involvement, many are struggling to find trustworthy partners. This raises questions as to whether clients and their consultants are utilising suitable pre-qualification selection criteria. Many clients undervalue the importance of robust pre-qualification methods or, in some instances, overlook them altogether. Huang and Wilkinson (2013) stressed the importance of what they refer to as 'prior learning' as a trust-building tool, especially for new relationships. Good-quality pre-qualification processes increase the likelihood of successful project outcomes; however, many prominent researchers in the field of pre-qualification (Holt et al. 1995) generally concurred that many practitioners carry out only subjective pre-qualification checks and some do not carry out any kind of formal review of contractors at all. Ng et al. (2002) explored and proposed a concept that would allow contractors' key performance indicator (KPI) data to be collected, verified and disseminated to clients via a web-based platform.

The study by Taylor (2017) carried out a questionnaire survey of 31 construction procurers, with the objective to assess common pre-qualification practices and the subsequent effects on project outcomes. Statistical analysis and tests were carried out which confirmed that robust pre-qualification practices do improve the chances of achieving successful project outcomes. However, the findings also showed that the full range of good-quality pre-qualification processes available to practitioners are not used consistently. Other findings included: practitioners frequently fail to follow up with contractor reference checks, do not always use objective assessments of contractor key performance indicators (KPIs) and the use of BIM improves collaboration.

9.3 Articulation of the Problem of Selecting Contracting Partners from the Perspective of Construction Professionals

There is a perception amongst private clients and procurers of large construction projects that the majority of contractors are still not truly embracing a culture of open and collaborative working. Furthermore, many clients believe that some opportunistic main contractors are using early involvement in projects as a means to force unfairly inflated prices and risk onto clients, effectively holding clients to ransom in the knowledge that they do not have the luxury of time to re-tender the works. Despite a succession of government and academic reports, anecdotal evidence would appear to indicate that, whilst the majority of private clients have accepted the need for, and the potential advantages of partnering and early contractor involvement, many are struggling to find trustworthy partners whom they are willing to give repeat contracts to. This raises questions as to whether clients are utilising suitable and ethical pre-qualification selection criteria. Turkish et al. (2008) suggested that many construction professionals undervalue the importance of robust pre-qualification methods or in some instances overlook them altogether. The effects of clients selecting the wrong contracting partners are that schemes can suffer from viability issues leading to large delays in the pre-construction phase or in some instances with projects being jettisoned. Some construction professionals have either incurred long implementation delays or have introduced a viability section in their planning processes. Some projects have come to fruition but concluded with the perceived non-delivery of the project objectives by either contracting party. Others have resulted in conflicts and disputes which represents a wasteful use of resources and an unnecessary burden to the industry's productivity problems (Latham 1994). If there are widely held perceptions that projects that have been negotiated and have had early involvement from main contractors are suffering from unfairly inflated prices, why is it that contractors are not all making high profit margins? In the past year, 'we have had a litany of poor results from major supply-chain players' (Farmer 2017), which culminated in Carillion, the second largest main contractor in the UK, going into liquidation after amassing huge debts of around £1.5bn in 2018 (BBC 2018). Could it therefore be construed that if construction professionals are indeed paying inflated prices and contractors' profit margins, there is a compelling problem with the performance of construction projects?

9.4 A Historical Perspective of the Problem

Tendering originated from pre-contract communication between architects and builders. By the end of the eighteenth century, architects' roles were consolidated: into construction designers and 'leaders' of project coalitions, hence establishing 'traditional procurement' (Holt et al. 1995). Traditional procurement generally refers to procurement routes that involve the splitting of the design phases from the construction phases. Design teams in this scenario are employed directly by construction professionals, to both design and administer construction contracts. Normally, upon completion of the design processes and after competitive tender, contractors are employed to execute the construction works. Hence, with the exception of the quality of the workmanship and adherence to the drawings and specifications, contractors bear no responsibility for the suitability of the designs. Historically this method is widely recognised as being adversarial in nature and prone to contractor claims. Furthermore, as contractors are effectively excluded from the design processes, projects can suffer from buildability issues, which can have a detrimental effect on time, quality and cost. A Government committee report by Simon (1944, cited in Holt et al. 1995) examined and recommended suitable procurement methods to facilitate urgently needed new social housing stock and public buildings such as schools and civic buildings on a scale never previously, or for that matter since, seen. Simon made some far-reaching observations and recommendations around selective tendering. He articulated if companies are shortlisted, based on the quality of the service or goods that they provide, rather than open tendering based on lowest price, it was much more likely to ensure that the products fulfill clients' needs, for a fair price. It is generally accepted that the Simon Committee initiated the move away from open tendering in lieu of encouraging negotiated procurement; therein allowing contractors to become involved with the project at the design stages (Holt et al. 1995). It is, however, widely accepted that the recommendations made by Simon were not implemented by either the incumbent or successive governments during the next 20 years (Costain 1964). In 1964 the government commissioned a further report when Sir Harold Emerson was asked by the Minister of Public Building and Works to consider practices adopted for the placing and management of construction work and, to make recommendations for promoting efficiency and economy within the industry (Holt et al. 1995). The report led to the formation of a committee led by Banwell (1964) after whom the report became to be known by. The recommendations put forward by Banwell echoed many of those made by Simon and whilst unlike the apathy that followed the publication of The Simon Report, some of the recommendations were implemented by the government; they were not wholly incorporated into government procurement methods and were largely ignored by the private sector. It was not until the 1990s following the release of the two seminal government reports by Latham (1994) and Egan (1998) (both of whom were calling for root and branch transformations) that the industry as a whole began to accept that there was a need for change towards collaborative procurement and working methods. However some 11 years after Egan made his recommendations, Wolstenholme (2009) who was commissioned by Constructing Excellence to review progress within the sector, found that whilst there had been some progress, many attempts to change had been superficial and most of Egan's targets had not been achieved (EC Harris LLP 2013). The Farmer Report (2016) that was commissioned by the government to review the construction labour model vis-a-vis skills shortages highlighted that there was still a 'collaboration problem' in the UK and its adversarial approach remains a salient issue which makes the industry resistant to change.

9.5 Risk Considerations

No construction project is risk free. Risk can be managed, minimized, shared, transferred or accepted. It cannot be ignored.

Latham (1994)

Whilst the subject of risk is to a large extent beyond the scope of this chapter, it is important for construction professionals to understand the dynamics and approach to risk management in construction procurement, as this can be a major factor in engendering collaborative professional relationships (McDonald 2013). Furthermore, 'risk' is an inherent factor of all construction projects, as they are often unique, constrained and are subject to the vagaries of human characteristics and shortcomings (APM 2006, p. 26). One of the major difficulties with risk management for construction professionals is deciding and importantly, agreeing, which party will be responsible for manging specific risks and perhaps more pertinently, bear the consequences if the risk manifests. The consensus is that ethically, the risks should be allocated to the party best able to control them and should be balanced proportionally to the extent of the stakeholder's interests in the project (Lehtiranta 2014). Unfortunately, opportunism can be prevalent in construction projects as parties often have different organisational objectives and targets. Traditionally opportunistic behaviour has been countered by utilising control-oriented mechanisms but this can be counter-productive to engendering collaborative relationships (Osipova and Eriksson 2013, pp. 391–392). This raises the importance of trust and ethics between the parties: 'Absence of trust and collaboration in risk management led to a low level of risk communication during the procurement phase' (Osipova and Eriksson 2011, p. 1154).

An interview of UK main contractors for a study by Laryea and Hughes (2008) found that most of the respondents were not using formal analytical risk probability models such as Monte-Carlo simulation, but rather relied on intuition and experience to identify risks and quantify the likelihood of occurrence. When asked about common causes of increased risk allowances, the respondents mentioned short tender durations, incomplete designs, buildability issues and clarity of tender documents. As contractors are generally employed early in two-stage tendering procurement methods and are tasked with managing or at the very least participating in the design development, these issues should be less prevalent (Naoum and Egbu 2016) hence this could indicate that the comments relate to other types of tendering arrangements.

The government's drive for increased use of Building Information Modelling (BIM) should also reduce issues that arise from buildability problems (Naoum and Egbu 2016, p. 325) and according to Constructing Excellence (BRE 2018) this is one of the major practices that engenders collaborative working. However, Vass and Gustavsson (2017) opined that the industry and particularly private clients has been slow to adopt the use of BIM. Research figures listed by Ghaffarianhoseinia et al. (2017, p. 1049) indicated that participation rates for use of BIM in UK construction in 2017 were 39%, and asserted that the public sector accounts for approximately 40% of construction output at this time. As the government has been a major proponent of BIM since 2011, and made its use mandatory on all public-funded works in 2016 (Lindkvist 2015), these statistics would appear to suggest that use of BIM in the private sector is low.

9.6 Benchmarking

We do not need to measure everything that matters; we only need to measure the things that matter.

Saad (2005, as cited in Raymond 2008)

For construction professionals to be able to undertake the selection processes for their consultants and main contractors, it is important for them to be able to benchmark each contender against certain criteria. This enables them to measure using such criteria and weightings each submission, based on what they believe is important for them and their projects.

According to McCabe (2001, p. 27), the following definition by McGeorge and Palmer (1997) provided the most succinct but complete description of the essential features of benchmarking:

A process of continuous improvement based on the comparison of an organisation's processes or products with those identified as best practice. The best practice comparison is used as a means of establishing achievable goals aimed at obtaining organisational superiority.

McGeorge and Palmer (1997, as cited in McCabe 2001)

Benchmarking, however, has many different variations and uses as a management concept, depending on whether companies are comparing themselves against their peers or just comparing competing contractors as part of a selection process (Cowper and Samuels 1996). With regard to the selection of suitable contractors: 'the comparison between one company and another may depend on performance benchmarking' (Luu et al. 2008, p. 759). Luu et al. (2008, p.760) stated that benchmarking is a comparison of the performance of external organisations to try to improve the organisation's own performance. Rigby et al. (2014) described benchmarking as an 'information-based approach to performance management'. Saad and Patel (2006) suggested that benchmarking is a process of 'measuring, comparing, learning and improving'.'

According to Constructing Excellence and BRE (2018), companies carrying out benchmarking exercises should avoid trying to measure too much initially and should firstly focus on the key salient performance issues.

Latham (1994) suggested that a national database be established (Sutton 2008) and Egan (1998) concurred: recommending benchmarking as a tool, both to help contractors improve their own performances and also as an approach to assist clients with contractor selection.

Benchmarks can also be used to allow procurement organisations or clients to assess the performance of construction firms. Here information can be accumulated either by clients solely, based on their experiences or it can be requested from suppliers to help clients draw a detailed picture of their suppliers' performance (Rigby et al. 2014, p. 787).

Constructing Excellence (2018) suggested that there are five key steps in the benchmarking process:

PLAN: Clearly establish what needs to be improved – make sure it is important to you and your customers – and determine the data collection methodology to be used including any key performance indicators (KPIs).

ANALYSIS: Gather the data and determine the current performance gap – against a competitor, the industry or internally – and identify the reasons for the difference.

ACTION: Develop and implement improvement plans and performance targets.

REVIEW: Monitor performance against the performance targets.

REPEAT: Repeat the whole process – benchmarking needs to become a habit if you are serious about improving your performance.

BRE (2018)

9.6.1 Key Performance Indicators

A Key Performance Indicator (KPI) is the measure of performance of an activity that is critical to the success of an organisation.

BRE (2018)

The precise history of performance measurement is not known but the practice was used by the emperors of the Wei Dynasty in the third century AD to benchmark its family members. It is thought that its first use in industry was possibly by Robert Owen in the early 1800s in the Scottish mill industry (Banner and Cooke 1984, p. 328). Dr W. Edward Deming who was credited as being instrumental in the post-war economic recovery in Japan was a prominent advocate of performance measurement. He was also a firm believer that focusing on the customer was the key to improving performance (McCabe 2001 p. 27).

Benchmarking is identifying 'best practice' used by others and comparing and contrasting the results of organisations using a set of predetermined key performance indicators (Fernie et al. 2006). Latham (1994) highlighted the benefits of benchmarking and suggested that national forum be established, and an ombudsman appointed to monitor the performance of the industry. Many of Latham's recommendations were reiterated by Egan (1998) but interestingly Egan also suggested that a score card system be adopted and the names of the best performers made public. A national KPI working group was established who set out a framework by which the industry could measure its performance (The KPI Working Group 2000):

> Sir John Egan's report, Rethinking Construction, challenged the industry to measure its performance over a range of its activities and to meet a set of ambitious improvement targets. This is the KPI Working Group's answer to that challenge. It sets out a comprehensive framework which construction enterprises can use to measure their performance against the rest of the industry, and has been designed to be used by organisations, large or small, specialist or supplier, designers or constructors...The KPI Pack presents the construction industry's range of performance by presenting organisations with a framework to benchmark activities both at a broad level, and at a level much closer to the 'coal face' – such as rectifying defects and meeting clients' expectations.
>
> *Modified from The KPI Working Group* (2000)

The BRE, through its 'Constructing Excellence' programme and in conjunction with Glenigan, collected data from the construction industry and allowed subscribing members to monitor their own performance against the industry average (BRE 2018). Glenigan had published annual construction performance reports; however in respect to performance criteria, the data is summarised and anonymous therefore reflected only the results of the industry averages. It appeared to have little to offer to the enquiring procurer but is aimed at providing contractors with industry average performance data from which to benchmark themselves against. However, studies by Ng et al. (2002), Beatham et al. (2004) and Rawlinson and Farrell (2010) suggested that the industry as a whole has also under-utilised the use of KPIs as an improvement tool. In response to the Latham and Egan Reports, the UK construction industry has developed its own set of KPIs. However, their effective use has been limited (Beatham et al. 2004, p. 93).

9.6.2 Constructionline

Constructionline was set up by the government in 1998, initially as an approved contractor data base for the Ministry of Defence contracts (Steele et al. 2003). This was in response to the recommendation made by Latham (1994), who articulated that duplicating pre-qualification practices was a wasteful practice (Beales 2008). The government put Constructionline up for sale in 2014 (Mann 2014), and it was purchased by Capita in January 2015 for £35 m (Dunton 2015). Capita had been operating the platform for a number of years under a concession agreement (Mann 2014). It seemed that whilst Constructionline has been widely used for public contracts, the private sector has not wholly embraced the platform, as is evident by a group of main contractors only agreeing to adopt its use in 2014 for subcontractor pre-qualification health and safety criteria some 16 years after it was established (Construction News 2014). Bouygues UK (2018) stated that 'We're proud to be part of an initiative that seeks to save suppliers in the industry more than £25 million a year.'

It is apparent that Constructionline reduced bureaucratic red tape and unnecessary costs incurred in the duplication of submitting documentary information, particularly for many small and medium enterprises (SMEs) when responding to pre-qualification questionnaires (PQQs) as part of the bidding process (Leflty 2014). It was, however, evident that it did not become the score card league table that Latham (1994) envisaged.

The application process uses the Publicly Available Specification 91 (PAS 91) questionnaire that was established by the British Standards Institute (BSI) in response to the government's desire to streamline construction procurement PQQs (Constructionline 2018). PAS 91 is mandated by the government; however, it is not classified as a British Standard. Reeve (2013) suggested that, whilst it would require lengthy consultations, if it was given full British Standard status and possibly even made into a specification, it would be more likely to be fully adopted by the industry as a whole.

9.6.3 Centralised KPI Sharing Systems

Ng et al. (2002) explored and proposed a concept that would allow contractors' KPI data to be collected, verified and disseminated to clients via a web-based platform. They argued that such a system would allow registered construction professionals to link individual contractor's performance results to its own goals and objectives and streamline the process of contractor selection.

The information would be collected from construction professionals (the appraisers) who would rate contractors' key performance criteria as highlighted in Figure 9.1. It is acknowledged that there could be potential difficulties in obtaining accurate and fair appraisals. To overcome such challenges, they proposed that before scores and rankings were published the data would be vetted and checked by an independent panel and that contractors would be given the opportunity to appeal the results. Ng et al. (2002) cited studies by Palaneeswaran and Kumaraswamy (2001) as having demonstrated how a similar system of Contractor Performance Assessment Reporting System (CPARS) had been implemented and had proved successful by Government clients in the US. However, Bradshaw and Chang (2013) asserted that the marking criteria of CPARS can be inconsistent. This was supported in the following quotation:

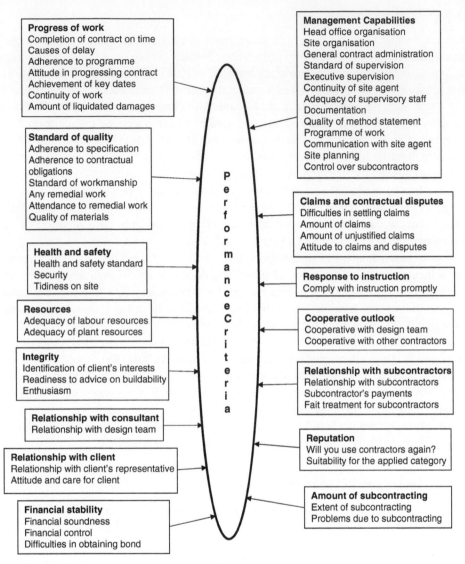

Figure 9.1 Performance criteria and sub-criteria.

Observations show that program managers and contracting officers are often reluctant to report negatively on past performance because this can reflect poorly on their own ability to manage the program or contract. Furthermore, to avoid conflict with the contractor, the government may refrain from documenting performance deficiencies in official databases. As a result, the past performance write-up does not always reflect a contractor's performance accurately.

Bradshaw and Chang (2013, p. 75)

It is somewhat unsurprising that contractors in the US have been known to litigate to contest appraisals by government contracting officers (Reed 2018), so this perhaps highlights the challenges of implementing a similar system for the private sector in the UK.

9.6.4 Common Pre-Qualification Practices

An effective selection process is crucial for clients wishing to strike a balance for successful project outcomes.

Fong and Choi (2000)

The Governments' Construction 2025 report (HM Government 2013) reinforced this view and advocates that selection processes should carefully evaluate contractors' experience, skills, resources and expertise rather than simply appointing contractors on lowest tender (Challender et al. 2014, p. 1041).

One of the most important activities for construction professionals is the selection of main contractors, and the use of robust pre-tender or pre-qualification selection methods which influence the likelihood of achieving clients' objectives and successful project outcomes (Hatush and Skitmore 1997b). Peace (2008, p.33) supported this view adding 'the selection and appointment of construction firms is one of the most important steps the client's internal team take to ensure a project's success'. Peace also stated that data collected during the evaluation phase should be assessed systemically and decisions taken on an objective basis. However, Deloi (2009, p. 1247) indicated that in the UK there was an overreliance of subjective analysis of whether contractors would likely be able to successfully deliver projects and meet client's objectives. This view was supported by Turskis (2008, p. 225) who argued that the selection of contractors is often based on intuition. Accordingly, a crucial task in contractor pre-qualification is to establish a set of decision criteria through which the capabilities of contractors are measured and judged. However, in the UK, there are no nation-wide standards or guidelines governing the selection of decision criteria for contractor pre-qualification (Ng and Skitmore 1999, p. 607).

Hatush and Skitmore (1997a), Holt et al. (1995), had identified common criteria for pre-qualification and bid evaluation and proposed improved methodologies for contractor selection. Holt et al. (1995) proposed a quantitative multi-attribute–based model, whereas Kumaraswamy (1996) discussed contractor evaluation through an appraisal of inputs and assessment of outputs using 'feedforward' and 'feedbackward' approaches. Hatush and Skitmore (1997) prescribed a (Program Evaluation and Review Technique) PERT-based methodology for assessing and evaluating contractor data for the purposes of pre-qualification and bid evaluation.

To maximise the chances of successful projects construction professionals should first consider and decide what represents quality to their organisations or to particular projects (APM 2006, pp. 22–23). This criterion can then be used to compile a weighted list of requirements from which contractors can be judged on their respective attributes against the set values and a scoring matrix (Thomas and Thomas 2005). However, Hatush and Skitmore (1997b, p. 129) argued that there is evidence to suggest that this approach is seldom adopted by most clients (and consultants) during pre-qualification selection processes. They concluded that many are 'more concerned with the process of retrieving completed proforma from candidate contractors than with under- taking any serious study of the relationships of this data to the project objectives'.

Ogunsemi and Aje (2006, p. 36) argued that there are five stages of pre-qualification criteria: 'references; reputation and past performance; financial stability; status of cur- rent work programmes; technical expertise; and project specific criteria'. In contrast, Hatush and Skitmore (1997a, p. 20) argued that the most common criteria considered by procurers during the pre-qualification and bid process are those pertaining to finan- cial soundness, technical ability, management capability and the health and safety per- formance of contractors. Gismondi (2017) highlighted the importance of following up with contractors' references and asking them to provide a comprehensive post-con- tract performance review of contractors. Gismondi also concluded that construction professionals should follow up references from contractors' past clients not listed as referees.

Although there appears to be an overwhelming consensus that good-quality pre- qualification processes have a positive correlation with improved project outcomes, a study by Wei et al. casts some doubt on this hypothesis:

> *It seems reasonable to expect that high prequalification requirements will result in a competent contractor and consequently deliver a quality product in a timely man- ner and within budget, the findings of this study suggest that this is not necessarily true.*

Wei et al. (1999)

Wei et al. (1999) suggested that the one reason for this could be that construction profes- sionals who possess comprehensive pre-qualification practices tend also to have bureau- cratic post-award practices. These could stifle projects, causing delays through over complicated processes to deal with issues such as change management.

9.7 Pre-Qualification Models and Methodologies

Ng and Skitmore (1999, p. 607) highlighted that despite Latham (1994) calling for stand- ardisation of pre-qualification procedures, no one single model has been accepted and is widely used by the industry and that this is probably due to differing construction client objectives. However, there are many different models available to aid clients with contractor selection decisions. These include multi-criteria model, fuzzy set model and the financial model (Darvish et al. 2009, p. 611).

9.7.1 Multi-Criteria Model

The Latham report (1994) questioned the practice of selecting contractors based on the lowest price and suggested that a better approach would be to compare contractors based upon a range of attributes and deliverables. The multi-criteria model was developed to address this and involved ranking and weighting a range of client requirements and expectations against which contractors are compared (Jennings and Holt 1998). Contractors are not solely judged on a single attribute, such as 'the financial offer' but several attributes are considered to arrive at a selection that represents the overall best value (Holt 1998, p. 154). Latham confirmed that clients should base their choice of contractor on a value for money basis with proper weighting of selection criteria for skill, experience and previous performance, rather than accepting the lowest tender (Holt et al. 1995). Construction professionals should be aware that multi-attribute decision-making is defined by processes that involve designing the best alternative or selecting the best one from a set of alternatives, that has the most attractive overall attributes, and that involves the selection of the optimal alternative (Turskis 2008, p. 225).

Morote and Francisco (2012, p. 9) were critical of the multi-attribute model claiming that it relies too much on the 'judgements and preferences of decision makers' and qualitative opinions and weightings which can be fuzzy and random. Some prominent researchers Ng and Skitmore (1999) and Hatush and Skitmore (1997b) et al. endeavoured to address this randomness. Ng and Skitmore (1999, p. 609) established 35 pre-qualification criteria derived from their studies and questionnaires of client-side stakeholders (architects, project managers, et al.) which were weighted based on a mean average of the respondents responses regarding the importance rating for each criterion. Hatush and Skitmore (1997c) utilised Programme Evaluation and Review Technique (PERT) to assign weightings to criteria based on opinions of experienced practitioners, whom they asked to rank the likelihood of each criterion affecting time, cost or quality.

9.7.2 Fuzzy Set Model

Fuzzy sets were introduced by Zadeh in 1965 as an extension of the classical notion of sets. Fuzzy set theory is a mathematical theory in which elements have degrees of membership within a certain interval (Nasab and Ghamsarian 2015).

Obtaining accurate and objective pre-qualification scores is often problematic as 'there are many inexact or qualitative criteria that are difficult to measure' (Yawe et al. 2005). Uncertainty and subjectivity in data collection during contractors' pre-qualification can result in ambiguous and obscure scoring as criterion such as reputation can be a grey area. Fuzzy set model uses fuzzy set theory to improve the objectiveness of the assessment and results (Lam et al. 2001). A fuzzy logic system can be described as sorting linguistic qualitative data in a non-linear arrangement to compute numerical scales and sets. Fuzzy logic systems generally have four parts: fuzzifier, rules, inference engine and defuzzifier.

9.7.3 Financial Model

One of the main checks that construction professionals and consultants frequently utilise is a financial review to assess the suitability and capacity of contractors to undertake their projects. This involves the collection and scrutiny of contractors' financial statements and other information to establish if they have the economic capacity to carry out the work. Failure to carry out such checks could result in a contractors experiencing financial constraints and struggling to fund the regular progress of the works, or worse, they could become insolvent during the project (Huang et al. 2013, p. 254). There are a plethora of checks that can be carried out to ascertain contractors' financial standings, including their ability to obtain surety bonds (Awad and Fayek 2012, p. 89), availability of working capital and documentary evidence of annual turnover levels (Deloi 2009, p. 1249).

The two main financial reports used to assess the financial position of a company are the income statement and the balance sheet. The income statement describes the operational process over a period of time (usually one year) and reveals sales (revenues), costs, expenses and profits or losses of a company over that time. The balance sheet is a snapshot of the financial position of a company at a particular point in time and describes its assets and liabilities. Net worth is the difference between what the company has and what it owes. Net worth is also helpful in determining the extent to which a company is leveraged. Leveraging refers to borrowing money to operate the company or to buy out shares in the company. Leverage measures indicate how much of the ability of the company to conduct business comes from borrowed funds (Pilateris and McCabe 2003, p. 488).

9.8 Designing a New Way for Construction Professionals to Select Their Contracting Partners

Earlier in this chapter we have already identified a number of practices that could help to improve the chances of construction professionals selecting reliable and proactive contracting partners. Conversely, a number of allegedly common practices were also reviewed that could be counter-productive to engendering collaborative relationships. The study by Taylor (2017) was conducted to identify and determine a new more appropriate method to collate, measure and analyse contractor data with a view to selecting construction partners. The study collected data from survey questionnaires and interviews based on the extent to which pre-qualification questionnaires are used and to assess project outcomes. This was to measure whether the quality of pre-qualification questionnaires which construction professionals use influences project success. Examples of the questions used to measure each of these two variables are illustrated below:

Pre-qualification questions
General question asking if detailed pre-qualification methods are carried out. Asks the respondent if the pre-qualification questionnaire includes a section relating to contractors' past performances. Asks if contractors' referees are contacted to corroborate contractors' past performances. Asks if the past clients of contractors that were listed as referees are contacted to provide feedback on contractor's past performances.

Asks if official platforms such as Constructionline are contacted to obtain contractors' KPIs.

Asks if the respondent or the respondent's company are ever themselves contacted by official platforms such as Constructionline to rate contractors' performances.

Asks if the respondent or the respondent's company are ever themselves contacted by other clients or consultants to rate contractors' performances.

Asks if curricula vitae (CVs) are requested from contractors for proposed staff.

Asks if interviews of the contractor's proposed staff are carried out at pre-qualification stage.

Asks if the results of KPIs are ranked and threshold criteria used to disqualify applicants.

Asks if professional pre-qualification programmes, software or models are utilised to compare and select contractors.

Asks if financial health checks are carried out on contractors at pre-qualification stage, such as the one by Dun and Bradstreet.

Asks if BIM is used on projects.

Project Outcome Questions

Asks if contractors' stage-two prices were generally fair and reasonable.

Asks if contractors completed projects on time.

Asks if contractors worked in an open and collaborative manner.

Asks if the contractors' stage-two prices were within the original budget allowance set by the client.

Asks if the project risks were shared fairly and appropriately, inferring between the contractor and the client.

Asks if a disproportionate amount of 'construction' risks were borne by the contractor. The inference in the speech marks is to differentiate between general developers' risks and risks relating to the actual construction phase. This could also be loosely categorised as risks after construction has commenced.

Similar to the last question, this question asks if a disproportionate amount of 'construction' risks were borne by the client.

Asks if existing contractors are invited to tender for future competitive two-stage tenders.

Asks if the quality of completed projects is generally finished to an acceptable standard.

Asks if repeat contracts were or likely to be negotiated with existing contractors.

Asks if the project out-turn costs are generally fair and reasonable.

Asks if clients paid a premium for deterioration in the skills base of contractors' staff.

Asks if the standard of health and safety management by the contractor was good.

Asks if the contractor's sustainability ethos was good.

Asks the respondent to rate the overall performance of contractors.

9.9   The Quality of Pre-Qualification Processes and Their Influence on Project Success

The overall results from the study by Taylor (2017) concluded a positive effect of 14% correlation between pre-qualification practices and successful project outcomes, which is deemed to be a significant influence. Notwithstanding this relationship, not all construction professionals carry out formal pre-qualification processes all of the time. Surprisingly only 8 of the 31 respondents, stated that their company 'always carries out a detailed pre-qualification process before shortlisting contractors for two-stage tenders'. This is quite an alarming revelation and something construction professionals should address in 'turning the tide' and deploying best practices to support successful project outcomes. Accordingly, it would appear that significant improvements could possibly be made to increase the likelihood and the success rate of projects, by practitioners simply employing and carrying out more robust pre-qualification methods. This position is supported by of the construction professionals who was interviewed for this book who asserted that:

> *The more checks you do the better at pre-qualification stage to determine if a contractor is the right company and has the right ethical credentials to be given the opportunity and responsibility to tender for and more importantly successfully build your project!*

A section in the pre-qualification questionnaires relating to past performance found a significant degree of correlation of 17% against project outcomes. However, it appeared that the information that contractors provide to demonstrate past performance is not always corroborated. From the findings, contractors' referees were frequently contacted (70%) to obtain firsthand feedback on performance but past clients (not listed as referees) were contacted to obtain firsthand feedback in just over half of cases (52%). A further test on 'contacting contractor's referees' and project outcomes revealed a correlation of 17% which again represents a significant influence. These correlation trends support the views of Gismondi (2017) who opined about the importance of following up with contractor's references. Even to most lay persons this must seem like a glaringly obvious and sensible course of action. How many people would invite a builder into their home, to carry out the most menial of projects, without first obtaining some kind of reference? Yet only 25% of the respondents stated that they 'always' check out contractors' references.

The study by Taylor (2017) also aimed to determine and analyse if current pre-qualification practices are sufficient to allow clients to identify and select collaborative contractors. Unfortunately, only 35% of respondents stated that contractors had mostly or always worked in a collaborative manner, with a modest positive correlation of 8% between the overall pre-qualification scores and collaborative contractors. This was not a glowing endorsement that current pre-qualification practices are sufficient to allow clients to identify and select collaborative contractors. The results of the some of the data led to a speculation that clients are overlooking this key soft skills attribute in their selection processes. Accordingly, construction professionals should reflect on including softer skills-based criteria and analysis as part of their pre-qualification selection processes.

One of the reasons that use of Building Information Modelling (BIM) was included in the questionnaire was due to the discovery during the literature review that according to Constructing Excellence (BRE 2018) this is one of the major practices to promote collaborative working. The results of the Taylor study appeared to corroborate this assertion showing a positive correlation (11%) between BIM measures influencing collaboration. Unfortunately, despite BRE and the government's drive to promote BIM, it is not widely used in the private sector, as the total score for BIM use in the sample group was only 57%.

It was interesting from Taylor's study that contractors who were deemed to be 'collaborative' were rewarded with repeat opportunities to tender for further work for construction professionals. There was a moderately strong 16% correlation between the two events. Conversely, it was disconcerting to observe a lack of any correlation coefficient between collaborative contractors and negotiated further work. It appeared that the construction client respondents did not show any inclination to reward such collaboration by negotiating future contracts. This could demonstrate an unnecessary burden of wasted resource and tendering costs for construction professionals, which could be regarded as a *raison d'etre* of the collaborative working agenda. For construction professionals this should be an area of careful reflection.

The third and final aim of Taylor's study was to determine and analyse the extent that construction professionals are using objective assessments of KPIs to shortlist contractors for tenders. It was discovered that most researchers in this field claim that many construction professionals used only intuition and subjective judgement to pre-qualify contractors for tenders. An interesting correlation was observed between the use of KPIs as a pre-qualification method and project outcomes. There was a moderately strong (13%) correlation indicating that the use of KPIs increases the likelihood of achieving fair project out-turn costs. However, the results of the survey sample indicated that many construction professionals could be failing to take advantage of this simple technique.

For the purposes of the assessment, the KPI category was classed as an objective process. It could, however, be construed that unless the threshold criteria have been prescribed as, say a company-wide diktat, that this process is still a form of subjective assessment.

9.10 Conclusions

The following objectives have been discussed in this chapter and a compilation of the findings from Taylor's study have been summarised.

9.10.1 Determination of Whether Quality of Pre-Qualification Processes by Construction Professionals' Influences Project Success

It has been from Taylor's study that pre-qualification processes for construction professionals do indeed influence project success. Accordingly, it follows that those construction professionals that utilise a range of good-quality diligent pre-qualification checks, have a greater chance that such diligence will yield positive results and projects completed in a manner that is considered as being successful. Despite this fact, many construction professionals are failing to employ and make full use of all the pre-qualification tools available to them. In some instances, organisations are not using them at all before shortlisting and appointing contractors for two-stage tenders, which could be affecting their chances of procuring successful projects.

Construction professionals should develop and implement project-specific and objective pre-qualification processes and ensure that any information submitted by contractors is validated by contacting cited referees, and if possible, other past clients not identified by contractors.

9.10.2 Determination of Whether Construction Professionals' Current Pre-Qualification Practices Are Sufficient to Identify and Select Collaborative Contractors

The findings of the survey undertaken as part of Taylor's study showed only a modest correlation between the pre-qualification processes and 'collaborative contractors'. However, previous literature referred to earlier in this chapter clearly suggested that 'prior learning' about contractors' backgrounds and their *modus operandi* improves the prospects of encountering or engendering trust-based collaborative relationships. Logically it should follow that to increase the likelihood of sourcing collaborative contractors, construction professionals should utilise the wide spectrum of checks available to them during the investigatory pre-qualification stage. However, the findings show that many construction professionals have failed to take full advantage of all available sources of information and tools, such as contacting past clients for references and the use of BIM.

It is clearly evident that the construction industry, certainly in the UK, still has an 'adversarial problem', and the findings Taylor's study to some extent corroborate this. Adversarial approaches among main contractors is seemingly still a prevalent issue in the industry. However, the question arises, of how much of this is due impart to the lack of incentivisation from construction professionals to contractors to be more collaborative; clients have still not fully embraced the concept of collaborative working themselves.

9.10.3 Determination of the Extent to Which Construction Professionals Are Using Objective Assessments of KPIs to Shortlist Contractors for Tenders

According to Taylor's study, pre-qualification processes increase the likelihood of achieving successful project outcomes. Findings corroborated and showed that simple ranking and comparing of contractors' KPIs (with minimum set prescribed threshold criteria) improved the likelihood of projects achieving fair out-turn costs. Disappointingly only approximately a third of the construction client respondents stated that they either mostly or always used KPIs which could represent a non-committal approach in this case.

9.10.4 Potential Recommendations and Drivers for Change and Improvement

The initial aim of Taylor's study was to establish what information and methods are available to construction professionals to help them to identify and select suitable contractors for long-term collaborative relationships, and in doing so improve the chances of successful project outcomes. Irrespective of which pre-qualification methods are employed, the findings of this study showed that there are many tools and processes available to construction professionals, and if used, increase the chances (regardless of the *modus operandi)* of achieving successful project outcomes. However, to improve the general standard and promote increased use, maybe the construction industry should develop and implement a National British Standard (NBS) for pre-qualification. This could be utilised by construction professionals and assist them to remove much of the subjectivity and vagaries which are associated with current practices. Possibly future studies could consider the introduction of an industry-wide contractor KPI data-sharing platform for construction professionals. It is acknowledged that this would require much research and consultation, especially regarding the obvious legal complexities involving defamation and associated litigation cases. However, the industry could reach a workable solution and implement a KPI data-sharing platform, surely nothing would make a greater impact and allow construction to finally shake off its adversarial nature and embrace a future where excellence of service is rewarded with increased business opportunities.

9.11 Summary

It is important for construction professionals to be aware and practice ethical processes around selection and appointment of their contracting partners. If processes and procedure leading to contract award do not have the right level of transparency and sense of impartiality, then this can have very damaging repercussions. Having the right consultants and main contractors on board, in terms of ethics and trust, is arguably one of the most important aspects for construction professionals. For this reason, it is critical for selection processes, especially when using collaborative procurement strategies, in choosing the most appropriate consultants and contractors to realise benefits of partnering through pro-activity, building team spirit, employing lateral thinking and exploring alternatives. Furthermore, the selection processes should carefully evaluate the consultants' and contractors' experience, skills, resources and expertise rather than simply appointing on lowest tender price. Selection criteria and competition should be based on life cycle value for money rather than simply reliant on initial capital costs. However, history has shown that the full range of good-quality pre-qualification processes available to practitioners are not used consistently. Furthermore, practitioners frequently fail to follow up with consultant and contractor reference checks and do not always use objective assessments of key performance indicators (KPIs) as benchmarking for their selection methods. The effects of clients selecting the wrong contracting partners are that schemes can suffer from viability issues leading to large delays in the pre-construction phase or in some instances with projects being jettisoned. Construction professionals should be able to undertake the selection processes for their consultants and main contractors, and it is important for them to be able to benchmark each contender against certain criteria. This enables them to measure using such criteria and weightings each submission, based on what they believe is important for them and their projects. Benchmarking is identifying 'best practice' used by others and comparing the results of organisations using a set of predetermined key performance indicators. To maximise the chances of successful projects, construction professionals should first consider and decide what represents quality to their organisations or to projects. This criterion can then be used to compile a weighted list of requirements from which contractors can be judged on their respective attributes against the set values and a scoring matrix. The more checks that are carried out at prequalification stage to determine if a contractor is a natural fit for a given project and has the right ethical credentials the greater the chances of generating successful construction outcomes.

References

APM (2006). *Association for Project Management. Body of Knowledge*, 5th ed. s.l... Association for Project Management, APM Publishing, High Wycombe, ISBN 1-903494-25-7.

Awad, A. and Fayek, A.R. (2012). A decision support system for contractor prequalification for surety bonding. *Automation in Construction* 21 (1): 89–98.

Banner, D. and Cooke, R. (1984). Ethical dilemmas in performance appraisal. *Journal of Business Ethics* 3 (4): 327–333.

BBC (2018). Carillion collapse raises job fears. [Online] Available at: http://www.bbc.co.uk/news/business-42687032 [Accessed 17 January 2018].

Beales, R. (2008). Call to make Constructionline mandatory. *Contract Journal*, 20 February, p. 3.

Beatham, S., Anumba, C., and Tony, T. (2004). KPIs: A critical appraisal of their use in construction. *An International Journal* 11 (1): 93–117.

Bouygues UK (2018). Main contractor zone. [Online] Available at: https://www.constructionline.co.uk/member-zone/main-contractors [Accessed 24 May 2018].

Bradshaw, J. and Chang, S. (2013). Past performance as an indicator of future performance: Selecting an industry partner to maximize the probability of program success. *Defense AR Journal* 20 (1): 59–80.

BRE (2018). Constructing Excellence – KPIs and Benchmarking. [Online] Available at: http://constructingexcellence.org.uk/kpis-and-benchmarking [Accessed 18 May 2018].

Challender, J., Farrell, P., and Sherratt, F. (2014). Partnering Practices: An investigation of influences on project success. [Online] Available at: http://www.arcom.ac.uk/-docs/proceedings/ar2014-1039-1048_Challender_Farrell_Sherratt.pdf [Accessed 17 January 2018].

Construction News (2014). Contractors pledge support for pre-qual register. Contruction News, 7 May.

Constructionline (2018). PAS 91. [Online] Available at: https://www.constructionline.co.uk/pas-91 [Accessed 25 May 2018].

Costain (1964). Banwell Committee (Report). [Online] Available at: http://hansard.millbanksystems.com/commons/1964/nov/09/banwell-committee-report [Accessed 22 January 2018].

Cowper, J. and Samuels, M. (1996). Performance benchmarking in the public sector. [Online] Available at: http://www.oecd.org/unitedkingdom/1902895.pdf [Accessed 11 May 2018].

Critchlow, J. (1998). *Making Partnering Work in the Construction Industry*. Oxford: Chandos Publishing Limited.

Darvish, M., Yasaei, M., and Saeedi, M. (2009). Application of the graph theory and matrix methods to contractor ranking. *International Journal of Project Management* 27 (6): 610–619.

Deloi (2009). Analysis of pre-qualification criteria in contractor selection and their impacts on project success. *Construction Management and Economics* 27 (12): 1245–1263.

Dunton, J. (2015). Capita buys Constructionline for £35m. [Online] Available at: https://www.building.co.uk/capita-buys-constructionline-for-£35m/5073467.article [Accessed 24 May 2018].

EC Harris LLP (2013). Supply chain analysis into the construction industry: A report for the construction industry strategy. [Online] Available at: https://www.gov.uk/government/publications/construction-industry-supply-chain-analysis [Accessed 21 May 2018].

Egan, J. (1998). *Rethinking Construction*. s.l... The Crown.

Egan, J. (2002). *Accelerating Change. Rethinking Construction*. London: Strategic Forum for Construction. London.

Farmer, M. (2016). *The Farmer Review of the UK Construction Labour Model. Modernise or Die*. s.l... Construction leadership council (CLC).

Farmer, M. (2017). Modernise or die: One year on. Building, 13 October, p. 33.

Fernie, S., Leiringer, R., and Thorpe, T. (2006). Change in construction: A critical perspective. *Building Research and Information* 34 (2): 91–103.

Fong, P. and Choi, S. (2000). Final contractor selection using the analytical hierarchy process. *Construction Management and Economics* 18 (5): 547–557.

Ghaffarianhoseinia, A. *et al.* (2017). Building Information Modelling (BIM) uptake: Clear benefits, understanding its implementation, risks and challenges. *Renewable and Sustainable Energy Reviews* 75: 1046–1053.

Gismondi, A. (2017). Challenges with prequalification process addressed during panel. Daily Commercial News, 16 June, pp. 1–2.

Gulf Construction: Al Hilal Publishing & Marketing Group (2008, December 1st). Two-stage tendering is a useful option. ProQuest, p. 1.

Hatush, Z. and Skitmore, M. (1997a). Criteria for contractor selection. *Construction Management and Economics* 15 (1): 19–38.

Hatush, Z. and Skitmore, M. (1997b). Evaluating contractor prequalification data: Selection criteria and project success factors. *Construction Management and Economics* 15 (2): 129–147.

Hatush, Z. and Skitmore, M. (1997c). Assessment and evaluation of contractor data against client goals using PERT approach. *Construction Management and Economics* 15 (5): 327–340.

Holt, G. (1998). Which contractor selection methodology? *International Journal of Project Management* 16 (3): 153–164.

Holt, G.D., Olomolalye, P., and Harris, F. (1995). A review of contractor selection practice in the u.K. construction industry. *Building and Environment* 30 (4): 553–561.

Huang, W. *et al.* (2013). Contractor financial prequalification using simulation method based on cash flow model. *Automation in Construction* 35 (11): 254–262.

Huang, Y. and Wilkinson, I. (2013). The dynamics and evolution of trust in business relationships. *Industrial Marketing Management* 42: 455–465.

Jennings, P. and Holt, G. (1998). Prequalification and multi-criteria selection: A measure of contractors' opinions. *Construction Management and Economics* 16 (6): 651–660.

Kwakye, A.A. (1991). *Fast Track Construction*. Ascot: The Chartered Institute of Building.

Lam, K.C. *et al.* (2001). A fuzzy neural network approach for contractor prequalification. *Construction Management and Economics* 19 (2): 175–188.

Laryea, S. and Hughes, W. (2008). How contractors price risk in bids: Theory and practice. *Construction Management and Economics* 26 (9): 911–924.

Latham, M. (1994). The Latham Report. [Online] Available at: http://constructingexcellence.org.uk/wp-content/uploads/2014/10/Constructing-the-team-The-Latham-Report.pdf [Accessed 29 November 2017].

Leflty, M. (2014). Too much information? One day the state might be selling the family data. The Independent, 25 July, p. 56.

Lehtiranta, L. (2014). Risk perceptions and approaches in mult-organizations. *International Journal of Project Management* 32: 640–652.

Lindkvist, C. (2015). Contextualizing learning approaches which shape BIM for maintenance. *Built Environment Project and Asset Management* 5 (3): 318–330.

Luu, V., Kim, S., and Huynh, T. (2008). Improving project management performance of large contractors using benchmarking approach. *International Journal of Project Management* 26: 758–769.

Mann, W. (2014). Government to sell off Constructionline. New Civil Engineer, 15 July.

McCabe, S. (2001). *Benchmarking in Construction*, 1st ed. London: Blackwell Science Ltd.

McDonald, T. (2013). Collaborative risk management. [Online] Available at: https://www.fgould.com/worldwide/articles/collaborative-risk-management [Accessed 22 May 2018].

Morote, A. and Francisco, R.-V. (2012). A fuzzy multi-criteria decision making model for construction contractor prequalification. *Automation in Construction* 25 (4): 8–19.

Naoum, S.G. and Egbu, C. (2016). Modern selection criteria for procurement methods in construction: A state-of-the-art literature review and a survey. *International Journal of Managing Projects in Business* 9 (2): 309–336.

Nasab, H.H. and Ghamsarian, M.M. (2015). A fuzzy multiple-criteria decision-making model for contractor prequalification. *Journal of Decisions System* 24 (4): 433–444.

Ng, S., Palaneeswaran, E., and Kumaraswamy, M. (2002). A dynamic e-Reporting system for contractor's performance appraisal. *Advances in Engineering Software* 33 (7): 339–349.

Ng, S. and Skitmore, R. (1999). Client and consultant perspectives of prequalification criteria. *Nuilding and Environment* 34 (5): 607–621.

Ogunsemi, D. and Aje, I.O. (2006). A model for contractors' selection in Nigeria. *Journal of Financial Management of Property and Construction* 11 (1): 33–34.

Osipova, E. and Eriksson, P. (2011). How procurement options influence risk management in construction projects. *Construction Management and Economics* 29 (11): 1149–1158.

Osipova, E. and Eriksson, P. (2013). Balancing control and flexibility in joint risk management: Esons learned from two construction projects. *International Journal of Project Management* 31: 391–399.

Palaneeswaran, E. and Kumaraswamy, M. (2001). Recent advances and proposed improvements in contractor prequalification methodologies. *Building and Environment* 36: 73–87.

Peace, S. (2008). *Partnering in Construction*. Salford: Construction Managers' Library.

Pilateris, P. and McCabe, B. (2003). Contractor financial evaluation model (CFEM). *Canadian Jornal of Civil Engineering* 30 (3): 487–499.

Procurement Group (1999). *Procurement Guidance No 4: Teamworking, Partnering and Incentives*. London: H M Treasury.

Rawlinson, F. and Farrell, P. (2010). UK construction industry site health and safety management: An examination of promotional web material as an indicator of current direction. *Construction Innovation* 10 (4): 435–446.

Raymond, J. (2008). Benchmarking in public procurement. *An International Journal* 15 (6): 782–793.

Reed, J. (2018). When should a contractor contest a CPARS rating? [Online] Available at: http://www.mwllegal.com/2018/01/18/when-should-a-contractor-contest-a-cpars-rating [Accessed 21 May 2018].

Reeve, P. (2013). It's time to standardise in construction. Construction News, 22 November.

Rigby, J., Dewick, P., Courtney, R., and Gee, S. (2014). Limits to the implementation of benchmarking through kpis in uk construction policy. *Public Management Review* 16 (6): 782–806.

Saad, M. and Patel, B. (2006). An investigation of supply chain performance measurement in the Indian automotive sector. *Benchmarking: An International Journal* 13 (1): 36–53.

Steele, A., Todd, S., and Sodhi, D. (2003). Constructionline: A review of current issues and future potential. *Structural Survey* 21 (1): 16–21.

Sutton (2008). Call to make Constructionline mandatory. Contract Journal, 20 February, p. 3.

Taylor, S. (2017). *Prequalification practices and project outcomes in connection with two stage tendering*. Unpublished MSc Construction Management dissertation. University of Bolton

The KPI Working Group (2000). KPI Report for The Minister for Construction. [Online] Available at: https://assets.publishing.service.gov.uk/government/uploads/system/uploads/attachment_data/file/16323/file16441.pdf [Accessed 21 May 2018].

Thomas, G. and Thomas, M. (2005). *Construction Partnering and Integrated Teamworking*, 1st ed. Pondicherry: Blackwell Publishing Ltd.

Turskis, Z. (2008). Multi-attribute contractors ranking method by applying ordering of feasible alternatives of solutions in terms of preferability technique. *Technological and Economic Development of Economy* 14 (2): 224–239.

Vass, S. and Gustavsson, T. (2017). Challenges when implementing BIM for industry change. *Construction Management and Economics* 35 (10): 597–610.

Wei, L., Krizek, R., and Hadavi, A. (1999). Effects of high prequalification requirements. *Construction Management and Economics* 17 (5): 603–612.

Wolstenholme, A. (2009). *Never Waste a Good Crisis: A Review of Progress Since Rethinking Construction and Thoughts for the Future*. Loughborough Univerity Report.

Yawe, L., Shouyu, C., and Xiangtian, N. (2005). Fuzzy pattern recognition approach to construction contractor selection. *Fuzzy Optimization and Decision Making* 4: 2.

10

Codes of Conduct for Professional Ethics

There is only one ethics, one set of rules of morality, one code: that of individual behaviour in which the same rules apply to everyone alike.

Peter Drucker

10.1 Introduction to the Chapter

This chapter of the book will articulate why there needs to be ethical principles for those professional working in the construction industry and will explain the importance of such principles. It will also highlight the challenges for ethical teaching given that there are no universal standards around the world and between different professional institutions on what constitutes good ethical practices.

The chapter will explain codes of conduct in regulating professional ethics, what these comprise and how they are applied to improve practices and behaviours. In this regard it will discuss the benefits of having codes of conduct and how they vary between different professional bodies and institutions. Penalties and sanction for non-adherence and breaches of professional codes of conduct will be covered in the chapter alongside implications for public trust and confidence for violations of ethical regulations. Alongside this, the importance of maintaining public trust will be discussed given that the construction industry has traditionally been regarded in poorly performing in the past with low levels of client satisfaction. Conversely the reputational benefits of strict compliance with institutional codes of conduct will be articulated and why these are important for maintaining professional standards.

Embedding codes of conduct and ethical behaviours and standards into organisational culture by professional bodies will be examined alongside the factors which are influential for promoting best practices. In this regard strategies and models will be identified and discussed how they are maintained within organisations. Finally, comparisons of many professional bodies will be made in there approaches to professional ethics, standards and behaviours and the differences between their codes of conduct.

10.2   Ethical Principles for Construction Professionals

Owing to increasing concerns in many high-profile cases including those previously referred to in Chapter 5 of this book, demonstrating dishonesty and corruption, it is important for construction professionals to commit to and encourage project teams to comply with sustainable ethical principles. Codes of ethics which have been introduced have provided an indicator that organisations and institutions take ethical principles seriously as they should outline expectations for all personnel with regard to ethical behaviour and intolerance of unethical practices (CMI 2013).

Relationships between construction clients and the professional consultants and contractors they appoint rely on professional ethics and trust especially since fee agreements cannot accurately specify all financial contract contingencies for possible additional services (Walker 2009). The main motivation for the public relying on members of professional bodies relates to them giving advice and practising in an ethical manner (RICS 2010). Accordingly, the RICS has developed 12 ethical principles to assist their members in maintaining professionalism and these relate to honesty, openness, transparency, accountability, objectivity, setting a good example, acting within one's own limitations and having the courage to make a stand. In order to maintain the integrity of the profession, members are expected to have full commitment to these values.

Arguably the main deficiencies of codes of ethics had emanated from the notion that there are no universal standards and accordingly they vary between countries and different sectors in the building industry. Boundaries and barriers created by fragmentation and differentiation within the construction sector have possibly deterred any common frameworks of professional ethics emerging in the past (Walker 2009). This is an area that demands more attention through multinational dialogue across all areas of the construction sector, to overcome.

10.3   Codes of Conduct to Regulate Professional Ethics

Members of professional bodies are bound by codes of ethics, sometimes referred to as ethical principles, to address the issue of non-ethical behaviour and to attempt to provide a context of governance (Liu et al. 2004). Members of such institutions are usually bound by codes of conduct in the way they practice, and the institutions reserve the rights to take action against members who breach rules and regulations laid down. A professional code of conduct could be described as a minimum level of behaviour which is expected of an individual member of a profession. Normally such behaviour relates to professional practice and compliance with institutional rules and regulate and the intention is to preserve the reputation and good name of a profession through self-governance.

Almost every profession has its codes of conduct to provide a framework for arriving at good ethical choices (Abdul Rahman et al. 2007). In the United Kingdom and indeed worldwide, construction industry professionals have qualifying and professional institutions and bodies that represent each discipline such as RIBA, CIAT, CMI, RICS, CIOB and many more. These institutions have strict charters, professional codes of conduct and ethics, rules and regulations relating to professional standards that individual members are required to follow and adhere to. These may differ in some details from one institute to another, but the majority of them do require members to act professionally and use and apply common ethical and moral principles for conducting business. Professional institutions who introduce and administer the codes of conduct for their members will normally enforce sanctions and disciplinary action on their members for non-compliance. If these individuals fall short of the required standards, they could be subject to a disciplinary committee who in some cases could expel them. For serious cases possibly involving illegal activity, e.g. fraud the institution, could report the perpetrators to the relevant authorities for further action.

Most Institutes and organisations have separated codes of ethics from codes of conducts, and some like Chartered Management Institute (CMI) defines its codes of ethics as 'a statement of the core values of an organisation and of the principles which guide the conduct and behaviour of the organisation and its employees in all their business activities' (CMI 2010). One of the reasons that code of professional conduct is applied is to promote good conduct and best practice (RIBA 2005).

It is normal for professions such as the Royal Institution of Chartered Surveyors to enforce mandatory continuous professional development (CPD) on their members to ensure they keep abreast of skills, education and regulations. These are linked to the professional institutions' ethical commitments to keep their members attuned to new developments in the sector, especially around their specialist areas of competence and experience. Some of the criticism of codes of conduct have in the past been voiced at the constraints and limits of the codes and the ability to manage behaviours, especially in the construction industry.

Technological and scientific advances in recent decades have led to continued requirements for the review, update and introduction of new codes of conduct by institutions and organisations. Such measures are required in order to respond to changes to aspects of human activities and environmental changes such as global warming and other environmental issues. The construction industry arguably has more relevant to environmental issues than those of other sectors, possibly due to the building process which have traditionally been heavy on carbon generation. In addition, in recent years there has been an increased focus on the construction life cycle costs of building, and so energy efficiency and use of renewable technologies has become more important.

It is important for construction professionals to be aware that adopting the aforementioned codes of conduct gains and maintains respectability and integrity for project teams, and their respective professional institutions and organisations (Vee and Skitmore 2003). Stewart (1995), however, was sceptical of such institutional codes of conduct and explained that these are merely guidelines for professionals to interpret as they wish and do not promote values, ethics and morality accordingly. Clearly corrupt behaviour is subject to more than just policing by professional institutions and in some cases can be deemed criminal offences. This being the case, there is a strong argument to suggest that the law, coupled with institutional sanctions may present an acceptable level of deterrent in that professionals will think twice before committing unethical behaviour if consequences are considered grave enough. Lui et al. (2004) argued, however, that transgression will still prevail if detection of breaches is considered unlikely or disciplinary measures imposed for breaches are regarded as insufficient or too lenient. Conversely, there is an argument that ethical codes of conduct should not be regarded negatively as a framework for punishing breaches but positively in assisting professionals in recognising their own moral parameters (Henry 1995). Accordingly, this raises the question of the role of codes of conduct and whether their purpose should be more closely linked to promoting compliance. This could involve motivating professionals to behave in an ethical manner as opposed to them be seen as frameworks to impose punishments for potential breaches.

Sometimes professional bodies may be accused of compromising governance matters linked with their own codes of conduct. The Independent Review of RICS Treasury Management Audit Issues (RICS 2021) found that the origins of what went wrong within the RICS between 2018 and 2019 laid in the governance architecture of the institution. A lack of clarity about the roles and responsibilities of the Boards, the senior leadership and the management left cracks within which the Chief Executive and his Chief Operating Officer had become used to operating with little effective scrutiny. The release of the report on 9th September 2021 culminated in senior leaders with the RICS stepping down including the Chief Executive, President, Interim Chair of Governing Council and Chair of the Management Board. The 467-page report concluded that 4 non-Executive Board members, who raised legitimate concerns that the audit had been suppressed, were wrongly dismissed from the Management Board and that sound governance principles were not followed.

10.4 Maintaining High Standards of Professional Conduct and Competence

When considering codes of conduct, one needs to determine not only the particulars relating to as particular code but the context of what the code is intended for and the capability to comply with it. Introducing codes of conduct, codes of ethics, rules and regulations by professional institutions is a strong message and indication to members on what is expected of them in terms of behaviour. It also sends a strong signal to other stakeholders that unethical practices will not be tolerated (Fewings 2009).

Codes of conduct developed by professional institutions are designed to provide a strict code of adherence to ethical values whereupon the interests of communities and clients take priority over self-interests of individuals belonging to those professions (Haralambos and Heald 1982). For this reason, the primary reason for having codes of conduct are to ensure and embed governance and regulation of professional ethics. This is an important focus for the book when considering that relationships between clients and their professional contributors has relied in the past on a significant level of trust and professional ethics (Walker 2009). Furthermore, the heart of best practice in construction management is the maintenance of high standards of professional conduct and competence, underpinned by the principles of honesty and integrity (CMI 2010). This view was reinforced by Vee and Skitmore (2003) and they articulated the importance of codes of conduct as the tool to enable such standards. They argued that construction professionals should have the fundamental commitment to professional conscience and professional competence. Professional competence in this context can be defined as the capability to perform the duties of one's profession generally, or to perform a particular professional task with skill of an acceptable quality (Vee and Skitmore 2003). Furthermore, professional competence is predicated on having the broad skills, attitude and knowledge to work in a specialised profession or area. Disciplinary knowledge and the application of concepts, processes and skills are required as the measure of professional competence in any particular field. Professional competency is one of the five fundamental principles of professional ethics along with integrity, confidentiality, professional behaviour and objectivity. Notwithstanding this, each of the professional institutions have their own lists of technical and professional competencies, which are closely aligned to their institutional rules and regulations.

Another important role for codes of conduct, according to Lere and Gaumnitz (2003) is to influence decision making. Notwithstanding this assertion, the problem has been to establish those codes of ethics which are most effective at promoting ethical decision making. This involves determining those codes of ethics that impact decision making to the effect that they change the values and beliefs of individuals. Furthermore, according to Inuwa et al. (2015) codes of conduct are to ensure clients' interests are properly cared for, whilst wider public interest is also registered and protected. The RICS have articulated that as client's expectations have changed, business practices need to change also to respond to these. This could mean that ethical issues need to feature highly on the list of organisational strategies and business drivers. Codes of ethics established by professional institutions such as the RICS can serve as 'checks and balances' for individual members to try to curb unethical or immoral behaviours. This could assist in reducing the undesirable practices and behaviours on a global level.

10.5 Governance and Enforcement of Professional Ethics through Codes of Conduct

In practice, as previously referred to, professional ethics are policed by a national or international body to ensure a minimum standard of practice for organisations to strictly adhere to. Such bodies are most frequently professional institutions that seek to ensure that the behaviour of construction professionals is controlled and monitored by a strict code of ethics, which they create and maintain (RICS 2001). These codes of ethics, frequently referred to as codes of conduct, are primarily to ensure that integrity and ethics are maintained at all times. Institutions such as the RICS and RIBA have royal charters which strictly set down rules and regulations relating to professional standards, moral and ethics which all members must comply with. Members of professional bodies are bound by codes of conduct to address the issue of non-ethical behaviour and to attempt to provide a context of governance (Liu et al. 2004). Members of such institutions are then bound by such codes of conduct in the way they practice, and the institutions reserve the rights to take action against members who breach rules and regulations laid down.

Professional ethics through the aforementioned codes of conduct set down acceptable norms and thresholds relating to standards of practice. This creates ethical codes, codes of conduct and sets of rules by which member individuals or organisations are regulated. Furthermore, measures are taken to give more transparency relating to investigation of what could be deemed as behaviour or actions of an unethical nature. This adds robustness and due diligence to the process and could have the effect of boosting public confidence. Codes of conduct can sometimes be synonymous with rules to deal with certain issues and particular situations. Examples could include members of a professional institution maintaining indemnity insurance and processes for reporting malpractice of fellow members. The institutional rules and regulations more often than not cover potential conflicts of interest at an individual or corporate level. This would normally apply where a person or organisation has obligations to more than one party on a commission or contract and from wit her a private or commercial standpoint there is a clash of interests. This has to be considered from the position of trust that lies with that particular individual or organisation and measures to ensure that others are not disadvantages by a lack of impartiality.

There can be deterrents to avoid or reduce unethical behaviours and adherence with codes of conduct and these can be intrinsic to individuals. However, professional institutions or organisations can also impose extrinsic deterrents to such behaviours through enforcement provisions, and these may be made compulsory or voluntary. Individuals will still decide themselves whether to comply with such codes and so they cannot be forced to comply with them. However, professional institutions or organisations can impose strict penalties for codes which are violated. The degree to which individuals view such penalties will affect the extent to which they comply. For instance, if individuals regard the penalties as having negative consequences for them, then they will be more likely to comply and the codes in such cases become compulsory for them. Examples of such penalties could be instant dismissal from an organisation or expulsion from a professional institution. Alternatively, the decision makers regard the penalties as less serious for them, then even the largest penalties may not make compliance compulsory.

It is important for construction professionals to be aware that adopting codes of conduct, gains and maintains respectability and integrity for project teams, and their respective professional institutions and organisations (Vee and Skitmore 2003). Stewart (1995), however, was sceptical of such institutional codes of conduct and explained that these are merely guidelines for professionals to interpret as they wish and do not promote values, ethics and morality accordingly. Clearly corrupt behaviour is subject to more than just policing by professional institutions and in some cases can be deemed criminal offences. This being the case, there is a strong argument to suggest that the law, coupled with institutional sanctions, may present an acceptable level of deterrent in that professionals will think twice before committing unethical behaviour if consequences are considered grave enough. Liu et al. (2004) argued, however, that transgression will still prevail if detection of breaches is considered unlikely or disciplinary measures imposed for breaches are regarded as insufficient or too lenient. Conversely, there is an argument that ethical codes of conduct should not be regarded negatively as a framework for punishing breaches but positively in assisting professionals in recognising their own moral parameters (Henry 1995). Accordingly, this raises the question of the role of codes of conduct and whether their purpose should be more closely linked to promoting compliance. This could involve motivating professionals to behave in an ethical manner as opposed to them be seen as frameworks to impose punishments for potential breaches.

10.6 Misconduct and the Reputation of Professions

Construction projects have traditionally had a reputation for poor performance, coupled with low levels of client satisfaction (Latham 1994). Poon (2003) argued that there is a specific lack of research into the ethical issues of construction management that could be resulting in less than satisfactory project outcomes. This is especially the case as professionals rely on both knowledge and ethical conduct for the general public to have confidence in them. For a professional institution to gain public confidence in this regard, maintaining ethical practices are very important, as they have a direct influence on quality of those services provided to clients and the public image and perception.

Where misconduct and unprofessional behaviours have been identified in the past, this has had negative implications on the way the construction industry has been viewed. In some cases, this has led to public concern and government attention. Conversely, where there is deemed to be a high level of ethical practices adopted and adhered to by a profession, then this would normally imply high performance and low client dissatisfaction levels. When specifically considering the surveying profession one needs to consider those services not specifically related to construction projects but the whole diverse range of property services across the spectrum across the life cycle of buildings. In the past the RICS has focused on trying to improve the reputation of the institution in this regard and distinguish it from other services such as estate agency. This was in a concerted effort to improve trust amongst the public in light of cases of some cases of professional misconduct. Notwithstanding this position, Poon (2003) argued that the focus for training surveyors has traditionally been linked to providing core skills and 'hands on' knowledge required to perform their services. Such priorities are not always focused on improving the quality of those services provided and ethical considerations in some cases has been regarded as a sub-strand to working practices rather than a foundation those services are built up on. Perhaps this explains why in the past there have been calls for reforms to training and education programmes for surveyors and how these may assist in improving quality levels.

10.7 Embedding Ethical Codes, Behaviours and Standards into Organisational Culture

Liu et al. (2004) argued that in the construction industry there has been a focus in the past on culturally related attributes of project teams and a general commitment to quality. Notwithstanding this premise, they concurred that there has been evidence to suggest that ethical behaviours and standards are responsible for successful project outcomes.

Professional ethics can be reinforced through professional standards of professional bodies and codes of practice to gain respectability and integrity. However, these codes and standards need to be embodied in management practices and roles and responsibilities assigned to ensure compliance across ethics programmes. As such these codes are insufficient in themselves to ensure ethical conduct and accordingly require to be complemented with organisational responsibilities and employer training programmes. Such positions as ethics officers could be instrumental in 'turning the tide' in bringing ethics and ethical values to the forefront of organisational values. Such interventions and measures could shape how organisations are structured and how they function and bring reputational benefits for them. A survey carried out by Vee and Skitmore (2003), of construction professionals in Australia, found that most companies (90%) had subscribed to a professional code of ethics, albeit only 45% had 'gone the extra mile' and committed to ethical codes of conduct. Most of the organisations and individual who contributed to the survey agreed that ethical practice is considered to be an important goal for their respective organisations. Furthermore, most of them reinforced the perspective that business ethics should be governed and driven forward by personal ethics.

Professional codes of practice and conduct are committed to by members of institutions as a prerequisite for being a member of that institution. According to Stewart (1995) they should, however, not be used to teach ethics, values or morality. They simply are there to lay down rules of conduct as guidelines for action. Enforcement of the codes can be an important factor, and if there are little or no consequences, then it is unlikely that individuals will always act in the way that a particular institution intends. However, when consequences are grave for a particular breach people will think twice before infringing ethical codes of conduct. A similar phenomenon is apparent when the risk of detection of breaches is low. If individuals believe they can breach ethical codes of conduct and get away with it, with little chance of being found out, then it is more likely that they will commit unethical practices. In this case it can create an environment of opportunism where ethical behaviours are heavily influenced by the organisation. In this way ethical practices and behaviours can be delineated by the climate and culture of the organisations and the boundaries of socially acceptable norms that they operate. Ethical codes do not in themselves always solve moral dilemmas or reduce instances of illegal practices. However, they work best where they are supported by mechanisms and structures that ensure review, adjudication, communication, oversight, review and enforcement.

Liu et al. (2004) created an organisational ethics model, particularly focused on the perspectives of surveyors and data obtained from Hong Kong Institute of Surveyors (HKIS). It analysed several factors under the categories of ethical climate, culture and codes to assess which of the factors were more influential and important in promoting ethical standards and behaviours than others. These are listed in Figure 10.1.

It was concluded from the research by Vee and Skitmore (2003) that from the perspective of surveyors the dominant factor in both the public and private sectors was Laws and Professional Codes. This was related to surveyors, as professional practitioners, being more conformant to professional institutional standards which act as behavioural norms for them to practice within. Furthermore, the study showed that ethical codes are more effectively implemented in the public sector, where financial and behavioural

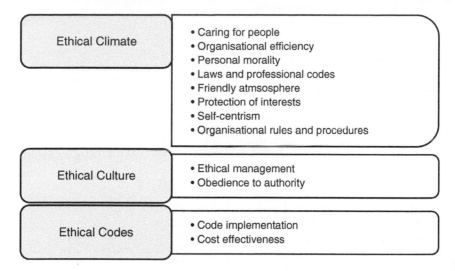

Figure 10.1 Model of ethical climate, culture and codes (adapted from Lui et al. 2004).

governance play a very important role in organisational management. These codes are deemed to be less clear in the private sector where their use in some cases could be more related to enhancement of reputation and image. For codes to be effective, it is crucial for them to be communicated down throughout the whole organisation.

10.8 Strategies for Improving Codes of Ethics Implementation in Construction Organisations

At an operational and corporate level within the construction industry there have been widespread and frequent problems in the past associated with business ethics. According to Kilcullen and Kooistra (as cited in Oladinrin and Man-Fong Ho 2014) business ethics are defined as 'a set of principles that guide business practices to reflect a concern for society as a whole while pursuing profits'. Accordingly, codes of ethics could be described as written organisational policy commitments to ethical conduct. Some examples of unethical practices which was covered in earlier chapters include mismanagement of client resources, poor quality of work and services, improper tender practices, bid cutting, collusion in tendering and improper relations with clients. In the past there have been attempts to explain how and why these practices and behaviours have been allowed to become commonplace. Possible reasons suggested by Oladinrin and Man-Fong Ho (2014) could include deficiencies in ethical codes of conduct embodiment within organisations and ineffective corporate code implementation. The presence of codes of ethics within organisations on their own are not sufficient to ensure that ethical practices and behaviours prosper.

One needs to consider how codes are implemented and thereafter how they are maintained within organisations, especially given that the global construction industry has been characterised in the past as having a poor ethical culture. Figure 10.2 illustrates some areas of implementation in this regard as devised by Svensson et al. (2009, as cited

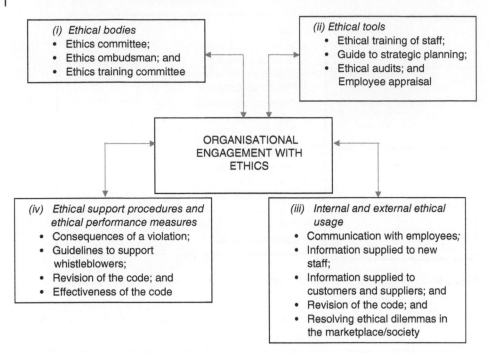

Figure 10.2 Areas of implementation of ethical codes (adapted from Svensson et al. 2009).

in Oladinrin and Man-Fong Ho 2014). These include ethical tools, ethical bodies, ethical support procedures and ethical usage.

These areas should be embedded into the structure and operation of organisations processes and procedures. Measures under each of these areas should include setting up of ethics committees and ombudsman, ethical training, conducting audits, communication with staff on ethical dilemmas and monitoring compliance with ethical codes. There have been other processes for implementation of ethics and the International Project Management Association (IPMA) have advocated and adopted the European Foundation for Quality Management Model (EFQM) in this regard. This is illustrated in Figure 10.3, and it is comprised of key component which are classified as enablers to implementing and embedding codes of ethics into organisations. Such enablers include processes, policy and strategy, leadership, partnerships and resources and people/employee management.

Each of these enablers would include six criterion each which represent attributes to stimulate ethical behaviours and improve ethical practices by embedding such codes. Leadership is regarded as being an important enabler as part of the EFQM model. Managers as leaders in their respective organisations are regarded as change agents, crucial to reshaping working practices and reforming cultures from the top down. This involvement of senior management is seen as paramount given that evidence has shown that ethical malpractice has resulted in the reputation of the construction industry being damaged. Accordingly, measures to correct this need to flow from top management with dedication and readiness to address breaches and instil an environment of integrity and due diligence. In addition, it is crucial to have the right and most

Hypothetical model for enabling codes implementation

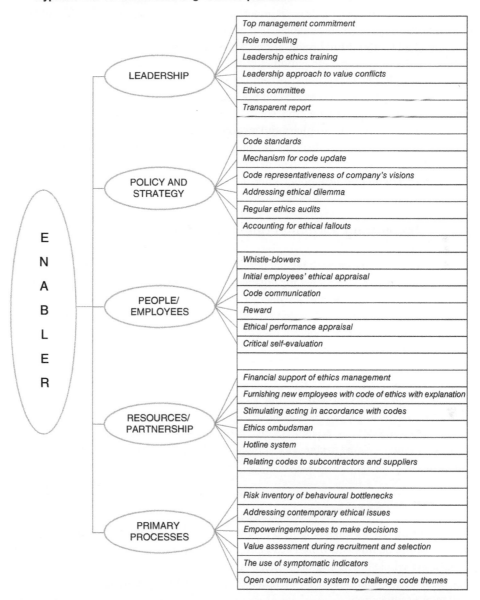

Figure 10.3 Enablers for implementation and embedding of codes of ethics (adapted from Svensson et al. 2009).

appropriate strategies and policies in place, as these are regarded as providing the means to deliver missions and visions on ethical reform. These should be supported by the relevant plans, targets, regulatory guidelines for implementing codes of ethics centred on stakeholders' requirements (Oladinrin and Man-Fong Ho 2014). It is accepted, however, that procedures, processes and regulatory guidelines on their own will not be sufficient without the commitment of employees. For this reason, the importance of human factors cannot be underplayed, especially as staff are positioned at the heart of any organisation and arguably their most important asset. According to Oladinrin and Man-Fong Ho (2014) is the 'people enabler' that provide the means for human resource management at all levels within organisations. Measures that could be implemented in this regard could include communication of codes, whistle-blowing protection procedures, appraisal of staff, reward systems and employee self-evaluation. The EFQM also recognises that managing ethics goes beyond the boundaries of organisations and extends to their stakeholders and wider supply chain. The importance of ensure the ethical compliance of subcontractors, suppliers and any other stakeholders that organisations contract within their business dealings is crucial. Accordingly, organisation need to be mindful how they approach the issue of their project participants and external partnerships in terms of their ethical policies and strategies.

10.9   Developing a Model Code of Conduct

Codes of conduct are sometimes anchored around the Seven Nolan Principles and embedded as core requirements for organisational financial regulations. According to Fewings (2009) these principles are detailed in Figure 10.4.

   The National Governance Association (NGA) have devised and advocate their Model Code of Conduct for Governing Boards (2019) which is based around the Seven Nolan Principles. They concur that:

> Effective boards set out clearly what they expect of individuals, particularly when they first join. A code of conduct should be maintained and communicated to all prospective appointees to set clear expectations of their roles and behavior. Explicit agreement to the code of conduct will mean there is a common reference point should any difficulties arise in the future.
>
> *The National Governance Association (NGA)*

   The NGA model is designed to help boards draft a code of conduct which sets out the purpose of the boards and which describes the appropriate relationship between individuals, the whole board and the leadership team of organisations. It should be noted that these codes of conduct can be tailored to reflect the specific governing board of particular organisations.

10.10   Chartered Management Institute (CMI) Codes of Ethics Checklist

A 'Code of Ethics Checklist' published by the Chartered Management Institute CMI sets down that ethics is particularly relevant to maintaining the reputation of an organisation and inspiring public confidence in it (CMI 2013). For this reason, codes of ethics

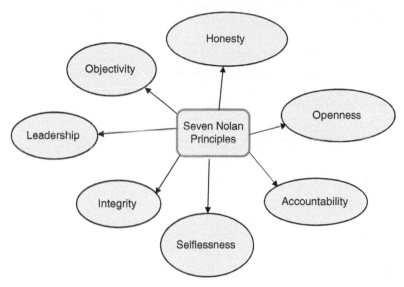

Figure 10.4 The Seven Nolan Principles. Fewings, P. 2009 / Taylor & Francis.

should reflect the practices and cultures which construction clients want to encourage for their respective organisations and project teams. This is supported by the CMI who advocated that:

> *A code of ethics is a statement of core values of an organisation and of the principles which guide the conduct and behaviour of the organisation and its employees in all their business activities.*

> CMI (2013)

In the context of some high-profile cases of unethical behaviour, sometimes involving fraud it is becoming ever increasingly important for firms to fulfill their commitments to ethical business principles. The Code of Ethics Checklist provides clear guidance for employees on what is expected of them in terms of ethical behaviours and practices. This sends a clear signal to other parties including customers and suppliers that unethical practices are not acceptable. The CMI articulates that a code of ethics must be more than just a document. They conclude that it should represent the culture and practice of an organisation and be arrived at through consultation and involvement with staff at all levels within an organisation. The action checklist is shown in Figure 10.5.

CMI advise that organisations should avoid ignoring the code once it has been introduced, so that it becomes worthless. They also concluded that it should not be used to impose inappropriate values and should not be regarded as a one-off rather than continuous commitment. They concur that it should not be a document that is confusing, vague, ambiguous and not comprehensible and in its creation should involve staff in the process of creating or revising the code.

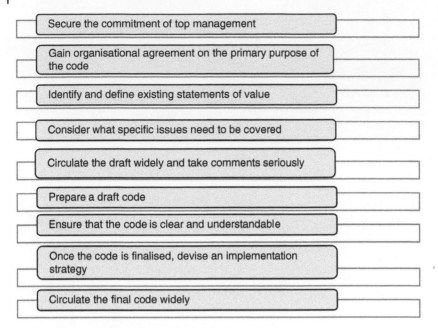

Figure 10.5 The Code of Ethics Checklist.

10.11   The Royal Institution of Chartered Surveyors (RICS) Codes of Conduct

The RICS have defined ethics as far as their institution around a set of moral principles and the notion of 'securing clients interest'. They have established an RICS Professional Ethics Working Party which emphasised the importance of ensuring that the interests of clients are adequately maintained and cared for whilst recognising and respecting the interest of the wider public (RICS 2000). The RICS administer the conduct of their members through the establishment of a set of Rules of Conduct which they stress all members must strictly adhere to, and these have over many years been regularly updated in line with the changing social environment. They cover such areas as conflicts of interest, professional indemnity insurance, professional and personal standards, rules on accounting, lifelong learning and disciplinary procedures. To assist in compliance, they have also published guidance documents with practical recommendations for setting up and governance of surveying practices, procedures for handling complaints, protection against money laundering and management of client confidentiality. The RICS have established nine core ethical principles in conjunction with the aforementioned Rules of Conduct and these are designed to assist members in certain challenging scenarios. These offer helpful guidance on ways and means to manage those difficult circumstances where members' professionalism may otherwise be compromised. In this way the RICS argue that there is less risk that members will breach ethical standards and therein improve the reputation of the institution and its members. The nine principles are illustrated in Figure 10.6. The nine RICS principles, coupled with the RICS Codes of

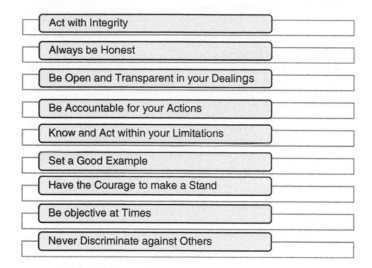

Act with Integrity

Always be Honest

Be Open and Transparent in your Dealings

Be Accountable for your Actions

Know and Act within your Limitations

Set a Good Example

Have the Courage to make a Stand

Be objective at Times

Never Discriminate against Others

Figure 10.6 The nine RICS ethical principles (adapted from RICS 2010).

Conduct, should together ensure that those professional services performed by surveyors are provided to their clients to maintain their interests at all times whilst preserving professionalism and the interests of the wider public.

The Royal Institution of Chartered Surveyors in their publication *Maintaining Professional and Ethical Standards* articulates that 'every member shall conduct themselves in a manner befitting membership of the RICS'. To support this, they advocate that 'members shall at all times should act with integrity and avoid conflicts of interest and avoid actions or situations that are inconsistent with their professional obligations'.

The RICS advocate that behaving ethically is at the heart of what it means to be a professional; it distinguishes professionals from others in the workplace. Furthermore, they require their members to demonstrate their commitment to ethical behaviour by adhering to five global professional and ethical standards and these are:

- Act with integrity. Be honest and straightforward in all that you do
- Always provide a high standard of service
- Act in a way that promotes trust in the profession
- Treat others with respect
- Take responsibility

The RICS have published regulations and guidance notes for clarity on each one of these standards and these are contained in Appendix I.

The RICS Codes of Conduct, in Table 10.1, are built up on the foundations of this commitment to professionalism and lists practical measures to ensure compliance in this regard. They deal with issues which include conflicts of interest, corruption, confidentiality, honesty and integrity. Members of the institution are then bound by such codes of conduct in the way they practice, and the institutions reserve the rights to take action against members who breach rules and regulations laid down. To put

Table 10.1 RICS Codes of Conduct.

RICS Codes of Conduct	Measures to Ensure Compliance
Act honourably	Never put your own gain above the welfare of your clients or others to whom you have a professional responsibility. Always consider the wider interests of society in your judgements.
Act with integrity	Be trustworthy in all that you do – never deliberately mislead, whether by withholding or distorting information.
Be open and transparent in your dealings	Share the full facts with your clients, making things as plain and intelligible as possible.
Be accountable for all your actions	Take full responsibility for your actions and don't blame others if things go wrong.
Know and act within your limitations	Be aware of the limits of your competence and don't be tempted to work beyond these. Never commit to more than you can deliver.
Be objective at all times	Give clear and appropriate advice. Never let sentiments or your own interests cloud your judgement.
Always treat others with respect	Never discriminate against others.
Set a good example	Remember that both your public and private behaviour could affect your own, RICS' and other members' reputations.
Have the courage to make a stand	Be prepared to act if you suspect a risk to safety or malpractice of any sort.
Comply with relevant laws and regulations	Avoid any action, illegal or litigious, that may bring the profession into disrepute.
Avoid conflicts of interest	Declare any potential conflicts of interest, personal or professional, to all relevant parties.
Respect confidentiality	Maintain the confidentiality of your clients' affairs. Never divulge information to others unless it is necessary.

this into context the RICS Professional Regulation and Consumer Protection Department have in the past dealt with approximately 21000 cases of professional misconduct mostly related to breaches in regulations, conflicts of interest and accounts breaches (RICS 2010).

To support their Codes of Conduct and Professional and Ethical Standards, the RICS have quite helpfully created a decision tree for proceeding or not proceeding with certain courses of action, and for decision making in these situations. This is contained in Figure 10.7 and carefully navigates members through potential ethical dilemmas that they may face in the course of their business dealings.

The RICS have also created a Frequently Asked Questions document linked to their Global Professional and Ethical Standards and this is contained in Appendix J. Furthermore, the RICS has regulations which sets down rules in the event that an organisation ceases to trade, in such circumstances related to maintaining professional indemnity, handling complaints and managing clients' monies. In the former case, professional indemnity 'run off' cover needs to be maintained for a minimum of six years following trading. This is to ensure that members and their organisations are not

Decision tree

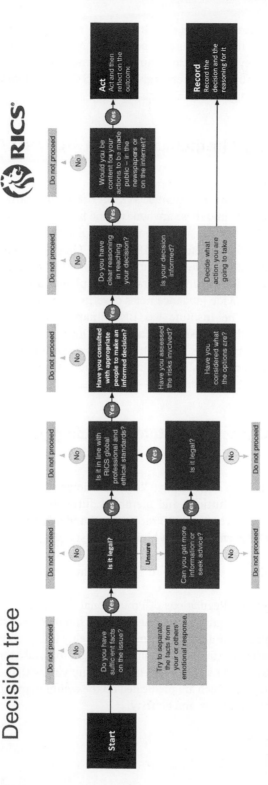

Figure 10.7 Decision tree for proceeding or not proceeding with certain courses of action (adapted from RICS 2010).

exposed to claims during this period. In the case of handling complaints, members should adopt an effective procedure for handling complaints prior to their organisation ceasing to trade. They should also ensure that clients' monies are preserved until all monies have been distributed. There is also the RICS Clients' Money Protection Scheme which provides robust governance in this regard.

10.12 The Royal Institution of British Architects Codes of Conduct

The Royal Institution of British Architects (RIBA) have three principles for their members to strictly adhere to and these are honesty and integrity, competency and relationship. These are built upon the institutional values of concern for others, environment, honesty, integrity and competency (RIBA 2005). These principles are contained in their Code of Conduct in Table 10.2 and are supported by clear rules and regulations around competition, advertising, insurance, complaints and dispute resolution. In addition, the RIBA provide guidance notes covering aspects as maintaining integrity and confidentiality. The Institute has Disciplinary Procedure Regulations that it administers, specifically related for breaches of ethical principles and regulations. It imposes disciplinary measures for contraventions according to the code and professional misconduct can be dealt with through sanction powers and a RIBA hearing panel.

10.13 The Chartered Institute of Building (CIOB) Codes of Conduct

The Chartered Institute of Building (CIOB) sets down its codes of professional conduct based on the institutions and members duties to clients, supply chain and employees. Furthermore, the CIOB recommend and advocate strongly around services needing to be performed to a high-quality standard and to achieve value for money for clients. They have also reinforced the importance of professionalism and duties owed to the general public to enhance the reputation of the construction industry.

The CIOB Rules and Regulations of Professional Competence and Conduct (CIOB 2019) consists of 14 rules governing what members should demonstrate, undertake and fulfill in discharging their duties with complete fidelity and probity. They also include four regulations related to use of distinguishing letters, logo, advisory service and advertising. Figure 10.8 contains some rules related to complying with their conduct for CIOB members in this regard which are predicated on the common premise of competencies, trust, respect, professionalism and honesty.

Table 10.2   Code of Conduct; The Royal Institution of British Architects.

1. **Principle 1 – Honesty and Integrity.**

1.1 The Royal Institute expects its Members to act with impartiality, responsibility and truthfulness at all times in their professional and business activities.

1.2 Members should not allow themselves to be improperly influenced either by their own, other's, self-interest.

1.3 Members should not be party to any statement which they know to be untrue, misleading, unfair to others or contrary to their own professional knowledge.

1.4 Members should avoid conflicts of interest. If a conflict arises, they should declare it to those parties affected and either remove its cause or withdraw from that situation.

1.5 Members should respect confidentiality and privacy of others.

1.6 Members should not offer or take bribes in connection with their professional work.

2. **Principle 2 – Competency**

2.1 Members are expected to apply high standards of skill, knowledge and care in all their work. They must apply their informed and impartial judgement in reaching any decisions, which may require members to balance differing and sometimes opposing demands (for example, the stakeholders' interests with the community's and project's capital costs with its overall performance).

2.2 Members should realistically appraise their ability to undertake and achieve any proposed work. They should also make their clients aware of the likelihood of achieving the client's requirements and aspirations. If members feel they are unable to comply, they should not quote for, or accept, the work.

2.3 Members should ensure that their terms of appointment, the scope of their work and the essential project requirements are clear and recorded in writing. They should explain to their clients the implication of any conditions of engagement and how their fees are to be calculated and charged. Members should maintain appropriate records throughout their engagement.

2.4 Members should keep their clients informed of the progress of a project and of the key decision made on behalf of the client.

2.5 Members are expected to use their best endeavours to meet the client's agreed time, cost and quality requirements for the project.

3. **Principle 3 – Relationships**

3.1 Members should respect the beliefs and opinions of other people, recognise social diversity and treat everyone fairly. They should also have a proper concern and due regard for the effect that their work may have on its users and the local community.

3.2 Members should be aware of the environmental impact of their work.

3.3 Members are expected to comply with good employment practice and the RIBA Employment Policy, in their capacity as an employer or an employee.

3.4 Where members are engaged in any form of competition to win work or awards, they should act fairly and honestly with potential clients and competitors. Any competition process in which they are participating must be known to be reasonable, transparent and impartial. If members find this not to be the case, they should endeavour to rectify the competition process or withdraw.

3.5 Members are expected to have in place (or have access to) effective procedures for dealing promptly and appropriately with disputes and complaints.

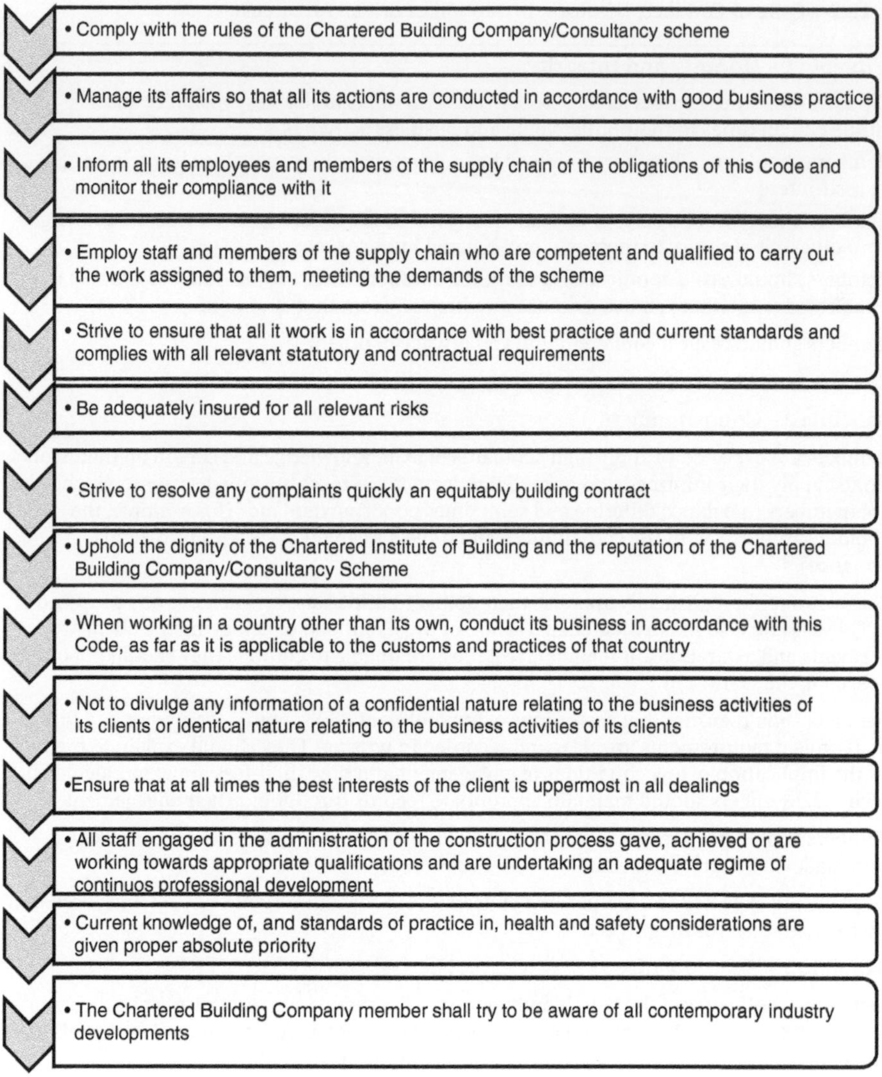

- Comply with the rules of the Chartered Building Company/Consultancy scheme

- Manage its affairs so that all its actions are conducted in accordance with good business practice

- Inform all its employees and members of the supply chain of the obligations of this Code and monitor their compliance with it

- Employ staff and members of the supply chain who are competent and qualified to carry out the work assigned to them, meeting the demands of the scheme

- Strive to ensure that all it work is in accordance with best practice and current standards and complies with all relevant statutory and contractual requirements

- Be adequately insured for all relevant risks

- Strive to resolve any complaints quickly an equitably building contract

- Uphold the dignity of the Chartered Institute of Building and the reputation of the Chartered Building Company/Consultancy Scheme

- When working in a country other than its own, conduct its business in accordance with this Code, as far as it is applicable to the customs and practices of that country

- Not to divulge any information of a confidential nature relating to the business activities of its clients or identical nature relating to the business activities of its clients

- Ensure that at all times the best interests of the client is uppermost in all dealings

- All staff engaged in the administration of the construction process gave, achieved or are working towards appropriate qualifications and are undertaking an adequate regime of continuos professional development

- Current knowledge of, and standards of practice in, health and safety considerations are given proper absolute priority

- The Chartered Building Company member shall try to be aware of all contemporary industry developments

Figure 10.8 Rules relating to the CIOB codes of conduct.

10.14 Summary

For a professional institution to gain public confidence, maintaining ethical practices are very important, as they have a direct influence on quality of those services provided to clients and the public image and perception. Codes of ethics can be described as a statement of the core values of organisations and of the principles which guide the conduct and behaviour of organisations and their employees in all their business activities.

Members of professional bodies are bound by these codes of ethics, sometimes referred to as ethical principles, to address the issue of non-ethical behaviour and to attempt to provide a context of governance. These institutions have strict charters, professional codes of conduct and ethics, rules and regulations relating to professional standards that individual members are required to follow and adhere to and they reserve the rights to take action against members who breach rules and regulations laid down. Accordingly, introducing codes of conduct, codes of ethics, rules and regulations by professional institutions instils a strong message and indication to members on what is expected of them and can serve as 'checks and balances' for individual members to try to curb unethical or immoral behaviours. They can also send a strong signal to other stakeholders that unethical practices will not be tolerated, and this could assist in reducing the undesirable practices and behaviours on a global level.

In considering the governance and regulation, professional ethics are policed by a national or international body to ensure a minimum standard of practice for organisations to strictly adhere to. Professional institutions or organisations can impose strict penalties for codes which are violated to deter unethical practices and behaviours. In this sense, however, ethical codes of conduct should not be regarded negatively as a framework for punishing breaches but positively in assisting professionals in recognising their own moral parameters. Clearly corrupt behaviour is subject to more than just policing by professional institutions and in some cases can be deemed criminal offences.

Where misconduct and unprofessional behaviours have been identified in the past, this has had negative implications on the way the construction industry has been viewed and, in some cases, this has led to public concern and government attention. To address this, there has been a concerted effort to improve confidence and trust amongst the general public in light of some cases of professional misconduct. Professional ethics, reinforced through professional standards of professional bodies and codes of practice, have played an important part in gaining this confidence and trust alongside respectability and integrity. However, these codes and standards need to be embodied in management practices and roles and responsibilities assigned to ensure compliance across ethics programmes.

Codes of conduct are sometimes anchored around the Seven Nolan Principles and embedded as core requirements for organisational financial regulations and these are selflessness, integrity, objectivity, accountability, openness, honesty and leadership. Codes of ethics should reflect the practices and cultures which construction clients want to encourage for their respective organisations and project teams. The Code of Ethics Checklist devised by the Chartered Management Institute provides clear guidance for employees on what is expected of them in terms of ethical behaviors and practices. This sends a clear signal to other parties including customers and suppliers that unethical practices are not acceptable. To support their Codes of Conduct and Professional and Ethical Standards, the RICS have quite helpfully created a decision tree for proceeding or not proceeding with certain courses of action, and for decision making in these situations.

References

Abdul Rahman, H., Karim, S., Danuri, M., Berawi, M., and Wen, Y. (2007). *Does professional ethics affect construction quality?* Retrieved from www.sciencedirect.com [accessed 24th November 2019]

CIOB. (2019). *Rules and Regulations of Professional Competence and Conduct.* The Chartered Institute of Building. Retrieved from www.ciob.org accessed 23rd May 2020.

CMI. (2010). *Code of professional management practice.* Retrieved October 12, 2010, from www.managers.org.uk/code/view-code-conduct [accessed 24th November 2019]

CMI. (2013). *Codes of Ethics Checklist.* London: Chartered Management Institute.

Fewings, P. (2009). *Ethics for the Built Environment.* London: Routledge.

Haralambos, M. and Heald, R.M. (1982). *Sociology: Themes and Perspective.* Slough: University Tutorial Press Limited.

Henry, C. (1995). Introduction to professional ethics. *Professional Ethics and Organizational Change*, 13.

Inuwa, I.I., Usman, N.D., and Dantong, J.S.D. (2015). The effects of unethical professional practice on construction projects performance in Nigeria. *African Journal of Applied Research (AJAR)* 1 (1): 72–88.

Latham, M. (1994). *Constructing the Team. London*: The Stationery Office.

Lere, J.C. and Gaumnitz B.R. (2003). The Impact of Codes of Ethics on Decision Making Some Insights from Information Economics. *Journal of Business Ethics* 48 (4): 365–379.

Liu, A.M.M., Fellows, R., and Nag, J. (2004). Surveyors' perspectives on ethics in organisational culture. *Engineering, Construction and Architectural Management* 11 (6): 438–449.

Oladinrin, T.O. and Man-Fong Ho, C. (2014). Strategies for improving codes of ethics implementation in construction organizations. *Project Management Journal* 45 (5): 15–26.

Poon, J. (2003). Professional ethics for surveyors and construction project performance: What we need to know. *Proceedings of The RICS Foundation Construction and Building Research Conference (COBRA) 1st September to 2nd September 2003.* London. RICS Publications

RIBA. (2005). *Code of Professional Conduct.* London: RIBA.

RICS. (2000). *Guidance Notes on Professional Ethics.* London: RICS Professional Ethics Working Party.

RICS. (2001). *Professional Regulations and Consumer Protection Department.* London: RICS House.

RICS. (2010). *Maintaining Professional and Ethical Standards.* London: RICS.

RICS. (2021). Independent review in respect of the issues raised at RICS in 2018 and 2019 following the commissioning by the RICS audit committee of a treasury management audit. RICS publications

Stewart, S. (1995). The ethics of values and the value of ethics: Should we be studying values in Hong Kong? In: *Whose Business Values*, (eds. S. Stewart and G. Donleavy), 1–18. Hong Kong: Hong Kong Press.

Svensson, G., Wood, G., and Callaghan, M. (2009). Cross-sector organizational engagement with ethics: A comparison between private sector companies and public sector entities of Sweden. *Corporate Governance* 9 (3): 283–297.

Vee, C. and Skitmore, M. (2003). Professional ethics in the construction industry. *Engineering,Construction and Architectural Management* 10 (2): 117–127.

Walker, A. (2009). *Project Management in Construction.* Oxford: Blackwell Publishing Ltd.

11

Implications in Practice for Ethics in the Construction Industry

The standard that a society should embody it is our own professed principles is a utopian one, in the sense that moral principles contradict the way things really are and always will be. How things really are and always will be is neither all evil nor all things good but deficient, inconsistent, inferior. Principles invite us to do something about the morass of contradictions in which we function morally. Principles invite us to clean up our act; to become intolerant of moral laxity and compromise and cowardice and the turning away from what is upsetting: that gnawing of the heart that tells us that what we are doing is not right, and so councils us that we'd be better off just not thinking about it.

Susan Sontag

11.1 Introduction

The above quotation clearly underpins the importance of ethical standards and principles in society. This chapter is largely focused on a research study conducted in Nigeria by Inuwa et al. (2015) and will consider the possible implications in practice when such standards and principles are not abided by or compromised in some way or form.

This chapter will explain the implications for the construction industry from unethical practices, which will include reputational damage and image considerations for organisations. In this regard it will report on high-profile cases that have received widespread press coverage and highlighted cartels, collusion on tenders, blacklisting of subcontractors and other illegal practices which have led to widescale public condemnation of the construction industry. It will then consider the effects on construction project performance and brought about by unethical practices, especially focused on those developing countries around the world with construction quality data generated from research studies in Nigeria. Thereafter, implications brought about by failures in construction quality will be discussed including future liability and vulnerability for organisations, increased maintenance costs, time overruns and reduced value for money, all being associated with poor performance. In addition, the possible remedies for poor quality and performance will be explored to counteract these dilemmas, including continuous professional development, engaging suitably certified construction professionals, transparency and accountability in contract administration and introducing strong

Professional Ethics in Construction and Engineering, First Edition. Jason Challender.

policy framework and enforcement strategies. Training and education around professional ethics will be discussed as a possible important remedy to 'turn the tide' on unethical practices, especially as the lack of suitable training and development for aspiring construction professional in developing countries still poses a major conundrum for the industry. In this respect the chapter will articulate how organisations should promote continued professional development in the workplace for the teaching and learning around ethics and the ways and means by which construction professionals can learn to adopt ethical behaviours and act with professional integrity in their practices and with a duty of care to their clients. Finally, the overriding need to embed more extensive and relevant training and education in curriculum at colleges and universities will be discussed and analysed.

11.2   Reputation and Image Implications for the Construction Industry on Unethical Practices

When unethical practices and behaviour come to light and especially when they come to light in the public domain, this can be extremely damaging not just for the individuals and organisation involved but for the reputation of the industry as a whole (Fewings 2009). Certainly, in the UK this has been a big issue for the construction industry over many years. As covered earlier in the book, reports and high-profile press coverage have highlighted cartels, collusion on tenders, blacklisting of subcontractors and other illegal practices this has led to widescale public condemnation of the industry. Whilst these reports and negative press coverage provide good news for newspapers, they destroy the image of the construction industry and present a very poor perception of it. Notwithstanding the fact that such practices may only represent a small minority of case, there is the perception that such behaviours are commonplace in the sector. To coin a phrase the construction industry and all the organisations employed within it are 'tarred by the same brush' by association. One could reach the conclusion that such widespread damage to the whole industry is unfair on those good organisations who are doing everything right to rebuild public confidence and act ethically. This has led to a poor image of the industry in the past and the perception that there are unregulated 'cowboy' builders operating widely and offering substandard services.

There are views that there are a few excellent contracting organisations providing outstanding services to their clients but many mediocre contractors who are still operating in a less than ethical way with poor working practices and behaviours. There have been reports that clients have been left with latent defects on work carried out, owing to poor quality of design, workmanship and materials. These problems may cause them long-standing difficulties in building performance and increases in maintenance and running cost in some instances. In extreme cases where contractors have refused to rectify faults and defects, clients have had no other option than to engage other contractors at additional cost to remedy the problems. In such cases this has also led to legal proceedings which have attracted negative publicity and further damaged the reputation and image of the construction industry.

11.3   The Effects on Construction Project Performance Brought About by Unethical Practices

According to Inuwa et al. (2015) there has been little research undertaken in the past relating to the effects of professional ethics on the performance of construction projects. In developing countries, there have been examples of corruption severely affecting the quality and in some cases compliance of buildings with current regulations. From the research study carried out in Nigeria by Inuwa et al. (2015), this has resulted in adverse events as show in Figure 11.1.

Furthermore, in extreme events this has resulted in the following outcomes as show in Figure 11.2.

From the same study by Inuwa et al. (2015), the research focused on the several identified effects of unethical practices from interviews with a whole range of clients, consultants and contractors. They offered some insight into the degree to which different unethical acts has on the performance and outcome of projects, according to the three classifications. Table 11.1 represents the findings in this regard through rankings for each unethical practice against clients, consultants and contractors. For clients the most

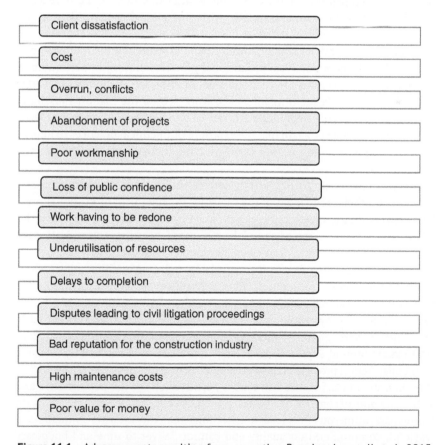

Figure 11.1   Adverse events resulting from corruption. Based on Inuwa, II et al., 2015.

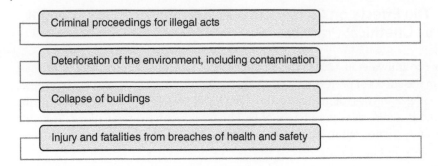

Figure 11.2 Extreme events resulting from corruption.

important negative effect on project performance was deemed to be linked to the bad image that unethical practices bring on the industry. Conversely, consultants felt that their most important issue was linked to unethical conduct leading to a future liability and vulnerability for increased maintenance cost associated with poor performance. Interesting main contractors ranked time overruns as being of the outmost importance when linked with unethical behaviour. This demonstrates that there is a difference of opinion on which negative outcome, associated with poor ethical conduct, depends on which role you have on projects.

When considering the combined ranking from clients, consultants and main contractors for each of the effects of unethical professional practices on project performance, Table 11.2 ranks those items from 1 to 20 in the order of importance and relevance. It is clear that vulnerability to frequent maintenance work scored the highest ranking of importance but perhaps surprisingly the high rate of site accidents scored the lowest ranking. This probably reflects, again that the exposure of unethical practices to longer-term maintenance liabilities and costs is a real consideration for clients, consultants and main contractors.

The research by Inuwa et al. (2015) also provided insight from clients, consultants and main contractors on their assessment of the remedies for unethical professional practices and which ones they felt were more relevant and important than others. Table 11.3 includes all the list of potential remedies and the rankings for the three classifications of roles. The need to adhere strictly to professional ethics was regarded by all the construction professionals as being a highly effective remedy for reducing or eliminating unethical professional practices. In addition, the adherence to project management methodologies was considered highly important by client with a ranking of 2 but of slightly lesser importance by consultants and contractors with rankings of 6 and 3, respectively. Furthermore, digital application in project management and elimination of dishonest claims around expertise, knowledge and skills was deemed by all three categories of construction professionals to be of relatively lesser importance.

It is clear than unethical practice can take a negative toll on the performance and ultimately the outcome of construction projects. Negative consequences associated with poor outcomes are numerous and lead to assertions that the industry is inefficient, unsafe, wasteful, compromised on quality and cannot deliver to time and budgetary constraints. This can severely impact on the reputation of the industry. There have been many ideas and opinions from clients, consultants and main contractors on the 'ways

Table 11.1 Ranking of the effects of unethical professional practices on project performance. Ranking1 = highest. Ranking 12 = lowest (adapted from data in Inuwa et al. 2015).

	Ranking in Terms of Importance/Relevance		
Effects	**Clients**	**Consultants**	**Main Contractors**
Abandonment	4	8	6
Clients/end-user dissatisfaction	10	10	10
Collapse of buildings	5	8	8
Conflict/disputes/litigation	9	9	6
Cost overrun	2	4	3
Delays	2	2	3
Deterioration of the environment	10	11	9
Deterioration of professionalism	4	6	4
High maintenance costs	4	8	7
High rates of site accidents	11	12	9
Poor aesthetic value	8	6	4
Poor basis for project monitoring and control	6	5	4
Poor clients' confidence on professional competence	8	5	5
Poor value for money	6	4	3
Poor workmanship	8	4	7
Portrays bad image of the construction industry	1	3	7
Rework owing to defects/snags	7	5	2
Time overrun	4	5	1
Underutilisation of resources	3	7	2
Vulnerability to frequent maintenance work	7	1	5

and means' through internal and external intervention mechanisms to curb such unethical practices. It is widely accepted that construction professional bodies, construction regulatory bodies and governments to adopt remedies, and such measures should always be embraced by construction professional in their respective roles.

Quality and ethics are inextricably linked through a common care premise to do the right things and therein a proven way to improve competitiveness, reduce costs and increase client satisfaction. Despite this, a study by Rahman et al. (2005) found that only a third of organisations surveyed had any ethical training programmes. Possibly this

Table 11.2 Combined ranking of the effects of unethical professional practices on project performance. Ranking 1 = highest. Ranking 20 = lowest (adapted from data in Inuwa et al. 2015).

Effects	Combined Ranking of Clients, Consultants and Main Contractors
Vulnerability to frequent maintenance work	1
Delays	2
Cost overrun	3
Time overrun	4
Portrays bad image of the construction industry	5
Underutilisation of resources	6
Deterioration of professionalism	7
Poor value for money	8
Poor basis for project monitoring and control	9
Abandonment	10
Collapse of buildings	11
Rework owing to defects/snags	12
Poor workmanship	13
High maintenance costs	14
Poor clients' confidence on professional competence	15
Poor aesthetic value	21
Conflict/disputes/litigation	17
Deterioration of the environment	18
Clients/end-user dissatisfaction	19
High rates of site accidents	20

emphasises that those that did not have programmes are unaware of the importance of professional ethics for their staff. Notwithstanding this premise, three quarters of those organisations in the study by Rahman et al. (2005) agreed that they felt the construction industry was tainted by unethical behaviours. One of the main examples of unethical conduct that the study revealed was associated with failures during bidding processes. Reports of lowest price awards, at the expense of quality, was commonplace. More serious examples of bribery and corruption were also reported.

11.4 Upholding Ethical Practices

In terms of teaching and upholding ethical practices, Helgadottir (2007) concluded that there are theories to consider, and these can be both process and outcome oriented. In the former case, ethics are considered from behaviours and actions are respectful to the duties and rights of others. In the latter case, ethics are viewed from the perspective that one's actions should lead to the best outcome for as many people as possible.

Table 11.3 Respondents ranking assessment of remedies for unethical professional practices. Rank 1 = highest. Rank 8 = lowest.

	Ranking in Terms of Importance/ Relevance		
Effects	**Clients**	**Consultants**	**Main Contractors**
Adherence to professional ethics	1	1	1
Adherence to project management methodology	2	6	3
Conduct project auditing using certified professionals	3	4	6
Continuous professional development	5	6	4
Elimination of dishonest practices and over-inflated claims around expertise, knowledge and skills	4	7	2
Engaging certified construction professionals	2	5	3
Digital application in project management	7	8	5
Legislate laws that spell out punishment for any type of unethical practice	2	4	2
Pre-emptive measures by regulatory bodies to supervise professionals	2	4	2
Strict disciplinary measures from property professionals	6	4	2
Strong policy framework and enforcement	2	2	2
The use of approved designs from certified professional	1	3	4
Transparency and accountability in contract administration	1	3	2

Accordingly, Helgadottir (2007) advocated that teaching students to analyse their own work from an ethical perspective is a valuable and effective exercise that students will find useful. Academic programmes should incorporate elements of ethical training and incorporate inventive ways and means to help students understand how to be competent in ethical issues. Organisations should look at their social responsibly in the context of managing projects.

11.5 Remedies for Unethical Behaviours and Practices in the Construction Industry

Unethical conducts can, especially in developing countries, undermine public confidence which can in turn hinder economic development, investment with a consequential loss of finance and in some cases human lives (Inuwa et al. 2015). For these reasons it has become an important priority for governments to address such practices in an effort to curb practices and cultures around unethical practices. This has put more centrally governed focus on creating policies designed to implement and police ethical reforms, and this could be through public procurement agencies. In Europe for instance the main public procurement agency is the European agency under the *OJEU* process that governs how the whole process of major projects and services over certain prescribed value limits are procured and this was covered in Chapter 6. These set down strict timelines and procedures of how public sector organisations can award contracts with heavy penalties for non-compliance and breaches of the rules. Some would argue that to truly embrace the dilemma of professional ethics, societies need to focus more on prevention and compliance rather than simply impose penalties for non-compliance. Such measures as including modules within university construction–related programmes could be beneficial to raise the importance of ethics in the professional arena. These could be formulated in a way to improve standards and create the right environment for motivating students to adopt and embed ethical standards when they graduate to professional practice. Professional institutions could also reinforce the same agenda through their codes of professional conduct and continuous professional development through conferencing and seminars.

11.5.1   Training and Education around Professional Ethics

A research study conducted by Rahman et al. (2010), in the context of developing economies, revealed that only 35% of respondents' construction organisations had conducted in any form of ethical training programmes. This low participation in training and education could possibly explain why the respondents felt that construction staff are not aware of the importance of ethics within their respective organisations. Accordingly, recommendations were made in the same study, that all construction projects should be managed by only those leaders who have received formal training in professional ethics. A similar assertion to support the findings from Inuwa et al. (2015) concluded that one of the contributory factors for unethical behaviours could be a lack of education and training of those individuals entering the industry. Accordingly, academic curriculum content relating specifically to the teaching of professional ethics has according to been severely limited in the past. This has been deemed to represent a gross inadequacy in the ethics education of potential construction professionals with the least covered area of ethics identified as corporate hospitality, receipt of gifts, entertainment, meals and whistle blowing (Inuwa et al. 2015).

   In consideration of the above, organisations should promote continued professional development in the workplace for the teaching and learning around ethics in practice to improve quality in construction outcomes and raise the reputation of the industry. In this way, construction professionals can learn to adopt ethical behaviours and act with professional integrity and a duty of care to their clients. Notwithstanding this premise, only when the practice of ethics is commonplace and placed centre stage within organisations will professionalism be enhanced leading to improved construction outcomes. Furthermore, parties within the construction industry should be vigilant, alert and proactive to avoid unethical behaviours and report any unethical practices of others to the relevant authorities. For this reason, the main public agencies, especially in developing countries, should adopt strict rules and regulations to enforce the values of ethical standards and outlaw those deemed by be unethical.

Mele (2008) argued that ethics should be integrated into the teaching of business management. This could take the form of inclusion of 'ethical reasoning skills' as the principal learning objectives for teaching programmes and the integration of ethical considerations into decision making. Possibly the four considerations for this would be to embed ethics into corporate governance, leadership and responsibility of business in society. However, the notion to integrate ethics into education can be challenging especially when considering that there are several different theories around ethics.

It has been widely documented that training and education in professional ethics in the construction industry are very significant in raising ethical standards and delivering more successful quality outcomes (Mohamad et al. 2015). Accordingly, the significance of education and continuous personal professional development around ethics is something that cannot be underestimated by professionals. Whilst most moral values are taught and learnt by individuals in early life there are many different reasons why people should continue to study ethics. There are many justifications for this position, and reasons have emerged that people require to know what to right from wrong to inform their judgements on what to do in certain situations.

Most members of society would agree that the competency, integrity and value of those who work in the construction industry are best developed firmly and deeply during their professional training and education whilst at college or university (Mohamad et al. 2015). Accordingly, some would argue therefore that there should be a more concerted effort to review courses and programmes at these educational establishments, to promote the study of ethics. This could take the form of looking critically at the role of construction professional in ethics awareness as part of their future leadership challenges. Professional decision making and actions should form part of such teaching and learning with ethics embedded in the business decision-making processes. Universities should in this sense be the pioneer for change and promote ethics to contribute and respond to the need for enhanced ethical practice. Notwithstanding this premises, most ethics-related courses or modules taught in colleges and universities have shown that they mainly form part of business-related courses. Courses and programmes linked to accountancy students have revealed that ethical considerations feature highly within them. Conversely those construction- and engineering-related courses appear to have lesser content and focus. The benefits of teaching ethics on these constructions- and engineering-focused course, according to Mohamad et al. (2015), could comprise:

- Allowing students to develop a set of moral values and to consider ethics in problem-solving solutions.
- Equipping students with an understanding of the basic construction-related issues and how these should be aligned and with moral and ethical practices.
- Encouraging students to develop critical judgements as opposed to simply taking on board and adopting the views and approaches of others, and in doing so reach their own answers.

Mohamad et al. (2015) argued that in consideration of the above, the educational context of universities presents an ideal environment to educate and practice ethics in terms of management and practice for the future. Furthermore, they opined that this is the environment that will assist in building tomorrow's construction community. Notwithstanding, these assertions, the learning of ethics should not cease for individuals in further or higher educations but should continue and be developed during their careers and professional working lives. Such individuals will possess different values of bad and good which in theory can result in a vast array of ethical or unethical practice. This can in most cases be promoted and governed by professional institutions, via Continued Professional Development (CPD) in their long-term role in ensuring professional standards of their respective institution. In this regard, such issues as maintenance and development of skills and knowledge, awareness of ethical considerations and maintaining public status all feature very highly. This is especially important within the context and environment of the built environment where things, regulations and developments are dynamic and constantly changing. The RICS approach this by imposing mandatory requirements for its members to submit written documented proof of at least 20 hours of CPD per calendar year, with penalties and sanctions applied for non-compliance. Some would argue that a 'carrot rather than the stick approach' would prove more successful and perhaps CPD participation in such areas of ethics could be better promoted through motivational measures rather than what could be regarded as old-fashioned penalties designed to punish. The implementation of CPD training may be best achieved through workshops, conferences, mentoring, lectures, webinars and by studying anew qualification. Other ways to disseminate CPD material could be through working forums, talks, discussions and seminars designing to endorse professional ethics.

11.6 Summary

There are many implications for the construction industry from unethical practices, which will include reputational damage and image considerations for organisations. Report of high-profile cases that have received widespread press coverage and highlighted cartels, collusion on tenders, blacklisting of subcontractors and other illegal practices which have led to widescale public condemnation of the construction industry. For this reason, when unethical practices and behaviour come to light and especially when they come to light in the public domain, this can be extremely damaging not just for the individuals and organisation involved but for the reputation of the industry as a whole. Notwithstanding the fact that such practices may only represent a small minority of case, there is the perception that such behaviours are commonplace in the sector,

It is clear than unethical practice can take a negative toll on the performance and ultimately the outcome of construction projects. Negative consequences associated with poor outcomes are numerous and lead to assertions that the industry is inefficient, unsafe, wasteful, compromised on quality and cannot deliver to time and budgetary constraints. Quality and ethics are inextricably linked through a common care premise to do the right things and therein a proven way to improve competitiveness, reduce costs and increase client satisfaction. From case studies in Nigeria, this chapter has highlighted many examples of unethical conduct associated with failures during bidding processes, with reports of lowest price awards, at the expense of quality being commonplace. Other more serious examples of bribery and corruption have also been covered, especially in developing countries, which has served to undermine public confidence and hinder economic development and capital investment. For these reasons it has become an important priority for governments to address such practices to curb practices and cultures around unethical practices. This has put more centrally governed focus on creating policies designed to implement and police ethical reforms, and this could be through public procurement agencies. Notwithstanding this premise, low participation in training and education, certainly in developing countries, could possibly explain why the respondents felt that construction staff are not aware of the importance of ethics within their respective organisations. Accordingly, one of the contributory factors for unethical behaviours could be a lack of education and training of those individuals entering the industry. To respond to this deficiency, organisations should promote continued professional development in the workplace for the teaching and learning around ethics in practice to improve quality in construction outcomes and raise the reputation of the industry. In addition, it has been suggested that colleges and universities should tailor more academic modules and courses to the subject of professional ethics in an attempt to address this deficiency in knowledge and education. Furthermore, organisations should be encouraged to instigate Continued professional development (CPD) of their employees to counteract unethical practices and behaviours within their companies. These organisations alongside public agencies and professional bodies, especially in developing countries, should also adopt strict rules and regulations to enforce the values of ethical standards and outlaw those deemed by be unethical.

References

Abdul-Rahman, H., Rahim, M., Hamid, M., and Zakaria, N. (2005). Beyond basic: The potential roles and involvement of the QS's in Public Projects-An Observation. QS National Convention 2005: Sustaining the Profession-Towards Diversification, 10–11 August 2005, Kula Lumpur, Malaysia:10 18.

Abdul-Rahman, H., Wang, C., and Yap, X.W. (2010). How professional ethics impact construction quality: Perception and evidence in a fast-developing economy. *Scientific Research and Essays* 5 (23): 3742–3749.

Fewings, P. (2009). *Ethics for the Built Environment*. London: Routledge.

Helgadottir, H. (2007). The ethical dimension of dimension of project management. *International Journal of Project Management* 26 (2008): 743–748.

Inuwa, I.I., Usman, N.D., and Dantong, J.S.D. (2015). The effects of unethical professional practice on construction projects performance in Nigeria. *African Journal of Applied Research (AJAR)* 1 (1): 72–88.

Mele, D. (2008). Integrating ethics into management. *Journal of Business Ethics* 78: 291–297.

Mohamad, N., Abdul Rahman, H., Usman, I.M., and Tawil, N.M. (2015). Ethics Education and Training for Construction Professionals in Malaysia. *Asian Social Science* 11 (4): 1911–2025.

12

Summary of Key Points, Reflections, Overview and Closing Remarks

If you're guided by a spirit of transparency, it forces you to operate with a spirit of ethics. Success comes from simplifying complex issues, address problems head on, be truthful and transparent. If you open yourself up to scrutiny, it forces you to a higher standard. I believe you should deliver on your promise. Promise responsibly.

Rodney David

12.1   Introduction

This final section of the book will summarise the book with reference to each, extrapolating the key findings and issues raised. Following on from what has been articulated and discussed, it will present some reflections and recommendations for the future of the construction industry, taking account of the inherent dilemmas for professional ethics.

12.2   Summary of the Key Issues Raised Throughout the Book

12.2.1   Application of Ethics in the Context of the Construction Industry

The construction industry accounts for a large percentage of the global economy and employs between 2% and 10% of the total workforce in most countries of the world. The performance of the industry can be seen to be a good indicator of the state and health of the wider economy. Despite the scale and importance of the UK construction industry, it is frequently criticised for performing poorly. This could be attributed to a general tendency for those employed in the construction industry to ignore best practice and be resistant to change.

   The construction industry has emerged as a bespoke project-based industry where there are many different characteristics to other industries such as manufacturing. This is largely due to construction projects being nearly always unique and 'tailor made' to suit clients' individual requirements. This bespoke project-based industry is made up of mostly small teams, ranging from construction workers to design consultants. Such teams come together on a temporary basis for the life of a project and then disband to undertake different projects and this makes it very different from most other industry

Professional Ethics in Construction and Engineering, First Edition. Jason Challender.

such as manufacturing. This also introduces fragmentation which can sometimes present an ever-increasing challenge for the sector. Owing to the fragmentation and bespoke nature of its composition, there are several unknown factors in any construction process which could cause the cost of projects to increase, programmes to be delayed and the quality of buildings to be compromised. This creates two aspects which come into play around commerciality and risk and who incurs any additional costs.

Project teams are made up of many construction professionals who have different roles and responsibilities. Clients are normally the project sponsors and accordingly they should be instrumental in steering, resourcing and leading all stages of the construction process from concept to completion. They appoint design teams which normally consist of project managers, architects, structural engineers, mechanical and electrical engineers and quantity surveyors. All these professional consultants work together closely within their different specialist areas during the design and construction phases. This coordinated and collaborative approach is designed to enhance teamwork for procuring successful project outcomes. On the contracting side of project team there are main contractors, subcontractors and suppliers. Main contractors are responsible for the day-to-day oversight of construction site activities which includes management of vendors and trades, and the communication of information to all involved parties throughout the course of building projects. Normally subcontractors are specialists in a particular area or trade, and they will perform all or part of the obligations of the main contractor's contract. It is common for a large construction project or renovation to involve many different subcontractors who work together to complete projects in a safe and timely manner.

The meaning of ethics can vary depending between the different roles and responsibilities of many different individuals that make up project teams. Clients for instance will expect ethical behaviours from their teams in managing projects closely and reporting any potential risks to them, as issues can frequently create additional cost and time delay implications for them. Conversely, from the consultant project team's perspective, there may be other ethical considerations to those of their clients. These may include expectations that clients are transparent in keeping them abreast of all issues especially around budgets and any issues that may affect progress of construction projects. From main contractors', subcontractors' and suppliers' perspectives, their view on ethics can be different again from clients and consultants. One of their main concerns is to be paid on time and to be treated fairly in commercial and contracting matters. Timely payments for these parties are especially important as cash flow for them is critical in their financial dealings.

There is an implied duty of care alongside established values and standards which govern whether decision making by construction professionals can be deemed good, bad or indifferent. However, there are sometimes problems for the construction industry in deciding how we judge what is right and wrong and therein what is acceptable. Models of investigation predicated on different branches of ethics can assist in deciding whether a particular situation or decision is ethical or unethical. These include the 'moral intensity', 'business', 'virtuous or professional' and 'integrity and reputational' models.

The life cycle of a building does not purely look at the build period, known commonly as the construction contract period or construction phase, but extends to the operational/

occupation period also, known as the 'in use phase'. There are competing interests at large between the design and construction phase and the in-use phase, normally predicated on capital costs versus operational costs. Decision making at the early stages of the life cycle can have long-lasting effects on the occupation in-use phase. Operational costs during the in-use phase normally represents most of the total cost of a building over its life whereas the capital costs during the development and construction phase representing only a fraction of total costs. Despite this, projects can become compromised by too much emphasis on the initial construction costs where capital savings may be required to meet budgetary requirements. These savings as sometimes referred to as 'value engineering' and this short-term planning can have long-lasting detrimental financial effects, which could far outweigh the original cost saving.

12.2.2 The Significance and Relevance of Ethics

There are alternative descriptions of what ethics are which highlights that the differentiation and variation around definition could give rise to problems of ambiguity and meaning. Accordingly, an ethical dilemma could ensue from the absence of any universally accepted definition of ethics. Perhaps this is one of the problems when considering professional ethics in the construction industry on a global scale, possibly leading to varying interpretations of ethics for construction professionals and what they mean in practice.

Goals of professional ethics can broadly be classified into two main categories: inward- and outward-facing goals. Inward-facing goals include creating common rules for people and organisations and outlining their responsibilities and duties to act and behave ethically. Conversely, outward-facing goals include providing a platform for evaluation of professions and as a basis for public expectations.

The importance of ethics should not be underestimated as they are considered a vital and essential practice requirement, engendering the trust of the general public and preserving their employers' interests. Accordingly, one of the benefits of ethics for organisations revolves around trust, especially where firms are heavily reliant on their reputation for conducting business and gaining new work. Furthermore, construction professionals who act in an ethical manner will enhance their performance which will increase the success of projects. Conversely, any compromises on ethics could jeopardise service delivery and damage public perception and the image of the construction industry.

It is important for construction professionals to commit to and encourage project teams to comply with sustainable ethical principles but there are many different aspects and issues that influence and affect professional ethics, and this presents a challenge for the industry. Codes of ethics which have been introduced have provided an indicator that organisations and institutions take ethical principles seriously as they should outline expectations for all personnel with regard to ethical behaviour and intolerance of unethical practices. Notwithstanding this premise, unethical behaviours and practices have been experienced in the construction industry over many years. There are repercussions that can arise from non-ethical practices, especially in the context of the UK construction industry. Construction professionals, as leaders in the procurement of projects, should be leading the way in cultural changes to improve the reputation of the industry. In this pursuit, they should be aware of the importance of ethics, alternative definitions and interpretations of ethics, the reputation of the construction industry, codes of conduct and governance and regulations in avoiding bad practices. In conclusion, measures to improve the practice of professional ethics, such as professional codes of conduct, have gone some way to improve the way the industry works but there are still far too many cases emerging of unethical practices that are blighting the sector. Although arguably these practices are emerging from a small minority of the sector, they are creating a bad press for the whole industry and further measures should be instigated by construction professionals to address this dilemma. Traditional responses in the past, at an institutional level, have been based on governance, regulations and punishment for non-compliance and clearly these have had only limited success. Perhaps construction professionals should be leading the way for a cultural change in the industry to train, educate and motivate construction individuals and organisations in what professional ethics entail, measures to ensure compliance and the benefits that they can bring for the sector. This could be achieved through more focus on FE, and HE course modules linked to professional ethics and CPD through workshops and training events in the workplace. These measures will hopefully contribute to providing a more ethical environment for the industry to work within and reap great benefits not just for all construction-related organisations and the building projects that they procure, but for the future of the construction industry at large. It is accepted, however, that to bring about these cultural changes will take conviction, integrity and in some cases courage not to engage in established unethical practices. These improvements once ingrained within the industry could then reap massive rewards in providing a safer, honest, trusting and more enjoyable working environment for all.

12.2.3 Ethical Dilemmas for Construction Practitioners

Construction-related ethical dilemmas can emanate from a variety of different issues that revolve around financial considerations, planning, conservation, waste management, land contamination, water, irrigation, pollution and energy. Professional ethics are not always aligned with business ethics, and there are sometimes conflicts with doing the right thing and making profits and achieving project targets. In such cases, financial considerations can become a pertinent factor in an imperfect environment where competing factors can create ethical challenges.

Few construction companies still do not include ethical issues as part of their strategic plans or mission statements. Furthermore, some organisations still have no ethical guidelines, policies or programmes for their staff to abide by. Possible improvement measures and reforms could come from business leaders in setting thresholds within their respective organisations for personal ethical standards which could be reflected and applied down the line to their employees. Compliance with such ethical standards maybe something that could be enforced through human resource management. This could take the form of training and education designed to make employees aware of what is acceptable and unacceptable practice and behaviours in the workplace. In this sense, professional conduct and ethical duties of individuals should be applied in the wider public interests and not only confined to the interests of clients and employers. Accordingly, construction professionals must not only take into account the interests of their clients and stakeholders, but the wider communities and the environment as well when taking actions and making decisions.

Adversarial attitudes in the construction industry have in the past affected relationships, behaviours, culture and trust in the construction industry. A major contributor in improving cultures with the construction industry has been professional ethics which defines rules of conduct. Notwithstanding these rules, ethical dilemmas and issues still frequently arise especially in the precontract stages and particularly during the tender processes for projects. With the award of construction contracts normally being synonymous with large sums of money and profits for the successful bidder, it is important to adopt best practices tendering procedures, compliant with the financial regulations of the awarding organisation. Any shortcomings in this regard could open the awarding body up for challenges from unsuccessful bidders especially where the awarding body is a public sector body. For this reason, as part of the tender evaluation processes there is a need to document each stage of the process, including the interviews and clarification sessions and treat all tender submissions equally. This documentation process is important as ethical dilemmas and issues around tender processes have in the past sometimes been challenged and escalated into court proceedings by unsuccessful bidders.

Most organisations normally have rules of governance and declaration procedures to cover corporate hospitality and gifts. However, there may be instances where the self-deception of an individual eludes themselves into the belief that a gift or corporate hospitality will not affect their judgement or impartiality in a particular case. To address this potential dilemma, most organisations have clear guidelines on giving and, perhaps more commonly, receiving gifts and hospitality for the benefit of their employees. In addition, some organisations have strict financial regulations which set down requirements for declaring or rejecting gifts and hospitality over a designated limit. The acceptability of the gift or corporate hospitality will depend very much on its value, but for modest infrequent gifts would normally be very much regarded as a 'token gesture' of goodwill. Unfortunately, there is no universal guidance on what is acceptable or unacceptable in the area of when gifts and hospitality become bribery. However, in deciding whether a gift should be accepted or rejected one needs to consider whether the offering could be considered to be excessive, be regarded as a norm for the business being conducted or whether it could influence unethical conduct or behaviour.

One of the main dilemmas for the community of professional practitioners and professional institutions is related to self-regulation and autonomy. This dilemma is accentuated by the different roles and responsibilities for the industry where diversity creates an environment of competing interest and conflicting ethical practices and standards. In addition, unethical behaviours and practices have in the past been attributed to the self-deception of individuals. In such cases, the self- deceiver's perception of things may encourage and enable them to act in immoral and unethical ways and deluding themselves that they are doing anything wrong in the process. Furthermore, there are scenarios where individuals are unaware that they have acted unethically and where their practices and decision making can be open to interpretation about what is right or wrong. Such dilemmas if unchecked could provide the platform for accountability, performance and quality to be compromised, to the detriment of clients and the general public. To address this conundrum, it is important for organisations and institutions to have robust regulation and governance of professional ethics with clear guidelines and rules. This could include whistle-blowing policies which have in the past addressed cases related to health and safety breaches, theft, bullying, cover ups, corruption and bribery. Such governance and regulatory policies and procedure are designed to give staff the ability to report such practices and therein provide a more transparent workplace, without the fears of harassment or victimisation.

12.2.4 Types and Examples of Unethical Conduct and Corruption

The construction industry provides a perfect environment for ethical dilemmas, with its low-price mentality, fierce competition and paper-thin margins. In addition, different cultures exist within the construction industry, and this may make boundaries between ethical and non-ethical behaviour become blurred at times. Categories of unethical behaviour could include breaches of confidence, conflicts of interest, fraudulent practices, deceit and trickery, presenting unrealistic promises, exaggerating expertise, concealing design and construction errors and overcharging. Less obvious examples of unethical behaviour could emanate from the misuse of power within organisations, and this could relate to managers making unfair or unreasonable demands on their subordinates, displaying intimidating behaviours or undermining their staff. There are degrees of understanding of what constitutes ethical compliance around regulations and rules, especially where tendering and procurement practices are concerned. Accordingly, one party may consider one practice to be acceptable and ethical, whereas another may find it totally unacceptable and unethical. This demonstrates that what constitutes unethical behaviour in practice is not always black and white and can come in many shades of grey owing to interpretation difficulties. This is particularly the case where gifts and corporate hospitality are concerned, albeit The Bribery Act 2010 does offer some clarity in the area; making it an offence to accept or give or receive in kind inducements to receive favours.

Some have argued that new players in the construction industry could be emerging with a general lack of appreciation of construction ethics. Notwithstanding this premise, causes of corruption can be exacerbated by factors such as greed, poverty, politics in the award of building contracts, professional indiscipline and favouritism. Corruption is particularly a problem in some developing countries where as much as 10% of total global construction expenditure was lost to some form of corruption. Notwithstanding this premise, corruption is not confined to developing countries and financial irregularities, scandals, blacklisting and what could the press have referred to as 'dirty tricks' have brought the UK construction industry into disrepute in recent years. Where corruption has occurred, it can be very damaging for the reputation of those organisations involved, especially where they have been found guilty of breaching ethical standards and committing criminal offences. Many cases of financial impropriety, false reporting and fraud have entered the courtroom sometimes resulting in fines, and in extreme cases imprisonment. The case of construction and professional services company Sweett Group PLC in February 2016 was one such example on a global scale, where they were ordered to pay £2.25 m after a bribery conviction. Other non-fine–related consequences for unethical breaches could include damage to reputation (personal and organisational), liberty (maximum 10 years prison sentence) and the inability to participate in tendering for public contracts. It is perhaps not surprising that the reputation of the construction industry has been tarnished over many years, which has culminated in a general lack of public trust and confidence in some countries.

Conflicts of interests have links with corruption and occur where there is a clash of commercial and private interests by individuals who hold a position of authority and trust. It could also be synonymous with an individual who has interests or obligations to two or more parties to a contract. Negligence in the design and construction of buildings represents another type of unethical action. Where contractors or designers have compromised the health, safety and well-being of individuals, including members of the public, through poor design or construction quality this could have catastrophic results with potential loss of life. Some examples of such cases include the partial collapse of Ronan Point tower block in 1968 and more recently the devastating fire spread at Grenfell Tower in 2017.

To reduce corruption across the world and restore trust in politics, it is recommended that global governments reinforce checks and balances around financial auditing, promote separation of powers, tackle preferential treatment, reduce biased towards special interests and prevent excessive money and influence in politics. Finally, other recommendations include managing conflicts of interest, promoting open and meaningful access to decision making and empowering citizens whilst protecting activists, whistleblowers and journalists.

12.2.5 Regulation and Governance of Ethical Standards and Expectations

Regulation of ethics should be focused on maintaining standards for all professionals practising within their respective areas of the construction industry. Such regulation should encompass expectations from the public and include responsibility, accountability, competence, honesty and willingness to serve the public. Notwithstanding this premise, there are dilemmas around how to regulate and govern such standards and expectations. Traditionally this has relied on self-regulation through professional institutions or governing bodies where behaviour of construction professionals has been controlled and monitored by strict code of conduct. However, in some cases, ethical quality control has been found to have fallen short in terms of projects not being completed on time and budget and to an acceptable level of quality, resulting in low client satisfaction levels.

Finance and governance regulations that organisations introduce provide the necessary framework for financial governance. The rules set down within these regulations should be operated transparently, consistently and designed to maintain the integrity of companies and their employees. In addition, they should be updated as necessary following consultation with internal and external stakeholders. Some organisations make it mandatory, when such changes are made for staff to undertake an induction to familiarise them with the new regulations, possibly via an online module. This is designed to ensure that they fully understand the new rules and regulations and it is not unusual for companies to make employees sign an undertaking to adhere to them. Financial governance policies and systems should be designed to address risks around ethical compliance. The pressures and temptations to cut ethical corners and to continue questionable practices can be a constant source of difficulty for companies. In this regard ethical risk could take the form of perceived pressure, perceived risk or rationalisation. In a concerted attempt to deter employees for breaching these ethical dilemmas, companies will normally impose rules and regulations related to purchase cards and raising purchase orders.

With regard to purchasing goods and services for companies, it is normal for thresholds to be created to mandate the procedures that need to be followed according to the value of those goods and services. In addition, companies normally regulate that all invoices require a purchase order to be raised in advance. This safeguards against individuals inadvertently verbally instructing work orders before such work orders are approved through the finance systems of their respective organisations. It is normal, certainly for public sector organisations, to have delegated authority limits for different individuals, boards and committees. These delegated authority thresholds are also a mechanism to prevent the large-scale abuse of budgets and expenditure and to deter unethical practices and corruption.

It is critical for employers to have ethically procedures in place during tendering and appointment stages for those contractors and consultants that they are planning to appoint. Maintaining records of each stage of the tendering process is very important especially when the awarding organisation is public sector, as challenges by unsuccessful bidders may necessitate documentation being presented in the public domain as part of the audit trail. By evaluating contractors and consultants on pre-determined criteria, it should allow employers to consider the quality of bids as well as the price. There is an ethical dimension to tendering, when contractors are required to contribute large sums of money and time to formulating bids that they stand little chance of winning. For this reason, it is both sensible and ethical to limit the number of tendering contractors to an appropriate level for them to invest the time and costs and stand a reasonable chance of success. Utilising recognised national procurement frameworks are regarded in the UK as a reputable and robust instrument for managing the award of contracts. Consultants and contractors on such frameworks are normally grouped into different categories according to the type a of work they undertake. In addition, for large public sector projects in Europe there are EU Procurement Directives which must be abided by. The most common directive for public sector procurement of goods, services, works and utilities is known as the *Official Journal of the European Union (OJEU)*.

12.2.6   Ethical Project Controls in Construction Management

The book has articulated the importance of construction professionals having robust financial processes in place. Accordingly, it is imperative that they understand their roles, responsibilities and authorisation levels when leading their project team. Furthermore, it is important for construction professionals to have robust decision-making processes in place within their respective organisations whilst ensuring governance procedures are maintained.

Project and programme boards are normally the vehicle by which the decision-making process is governed. Their main purpose is to provide governance and transparency to projects and avoiding the responsibility and accountability for decision making falling to one individual. With this in mind, where the value of a decision-making process exceeds the authorisation level of the project board then the decision making needs to be referred to a higher level. Accordingly, construction professionals need to be mindful when their projects require a higher tier of approval within their respective organisations, that they allow sufficient time in their project programme for the approval process.

In considering the approval processes for projects through the individual boards, it is commonplace for robust and financially rigorous business cases to be prepared to support the business venture. In some organisations, mostly public sector, there are gateway processes for project approval and business cases. The gateway processes will highlight different levels of approval for projects depending on which design stage they are at. Practical examples and templates of gateway applications, supported by business cases, have been presented in the book, to assist readers in this regard.

12.2.7 Developing an Ethics Toolkit, as a Practical Guide for Managing Projects

In general terms the information included in this book could become a useful, simple and innovative practical guide for construction professionals, especially those with little or no previous experience in managing projects. It could offer them a useful management tool, as relative beginners to the task of carrying out complex building projects, and hopefully narrows the gap between their existing knowledge base and those skills required for successful construction management. The various pro forma and templates, tailored to what is believed construction professionals require, should form the impetus for improved practices as checklists to ensure critical information and processes for ethical and legal compliance are strictly adhered to.

The opportunities for the guidance contained in this book could support future development for professional practice and in education context a useful teaching and learning resource or handbook. This is it presents a more effective and efficient means of collating building information and a step-by-step route map through complex procurement stages and processes. It has the potential to become integrated into BIM systems which client organisations may wish to implement. This technology could be networked and downloaded on to a software system that could effectively prepare a report automatically. This would make the whole process so much more efficient and greatly reduce the resources and time currently expended by construction professionals' staff in preparing various reports. This is particularly relevant for client companies seeking alternative ways to reduce costs and become more competitive accordingly.

12.2.8 Ethical Selection and Appointment Processes for the Construction Industry

It is importance for construction professionals to be aware and practice ethical processes around selection and appointment of their contracting partners. If processes and procedure leading to contract award do not have the right level of transparency and sense of impartiality, then this can have very damaging repercussions. Having the right consultants and main contractors on board, in terms of ethics and trust, is arguably one of the most important aspects for construction professionals. For this reason, it is critical for selection processes, especially when using collaborative procurement strategies, in choosing the most appropriate consultants and contractors to realise benefits of partnering through pro-activity, building team spirit, employing lateral thinking and exploring alternatives. Furthermore, the selection processes should carefully evaluate the consultants' and contractors' experience, skills, resources, and expertise rather than simply appointing on lowest tender price. Selection criteria and competition should be based on life cycle value for money rather than simply reliant on initial capital costs. However, history has shown that the full range of good-quality pre-qualification processes available to practitioners are not used consistently. Furthermore, practitioners frequently fail to follow up with consultant and contractor reference checks and do not always use objective assessments of key performance indicators (KPIs) as benchmarking for their selection methods. The effects of clients selecting the wrong contracting partners are that schemes can suffer from viability issues leading to large delays in the pre-construction phase or in some instances with projects being jettisoned. Construction professionals should be able to undertake the selection processes for their consultants and main contractors, and it is important for them to be able to benchmark each contender against certain criteria. This enables them to measure using such criteria and weightings each submission, based on what they believe is important for them and their projects. Benchmarking is identifying 'best practice' used by others and comparing the results of organisations using a set of predetermined key performance indicators. To maximise the chances of successful projects, construction professionals should first consider and decide what represents quality to their organisations or to projects. This criterion can then be used to compile a weighted list of requirements from which contractors can be judged on their respective attributes against the set values and a scoring matrix. The more checks that are carried out at pre-qualification stage to determine if a contractor is a natural fit for a given project and has the right ethical credentials the greater the chances of generating successful construction outcomes.

12.2.9 Codes of Conduct for Professional Ethics

For a professional institution to gain public confidence, maintaining ethical practices are very important, as they have a direct influence on quality of those services provided to clients and the public image and perception. Codes of ethics can be described as a statement of the core values of organisations and of the principles which guide the conduct and behaviour of organisations and their employees in all their business activities. Members of professional bodies are bound by these codes of ethics, sometimes referred to as ethical principles, to address the issue of non-ethical behaviour and to attempt to provide a context of governance. These institutions have strict charters, professional codes of conduct and ethics, rules and regulations relating to professional standards that individual members are required to follow and adhere to and they reserve the rights to take action against members who breach rules and regulations laid down. Accordingly, introducing codes of conduct, codes of ethics, rules and regulations by professional institutions instils a strong message and indication to members on what is expected of them and can serve as 'checks and balances' for individual members to try to curb unethical or immoral behaviours. They can also send a strong signal to other stakeholders that unethical practices will not be tolerated, and this could assist in reducing the undesirable practices and behaviours on a global level.

In considering the governance and regulation, professional ethics are policed by a national or international body to ensure a minimum standard of practice for organisations to strictly adhere to. Professional institutions or organisations can impose strict penalties for codes which are violated to deter unethical practices and behaviours. In this sense, however, ethical codes of conduct should not be regarded negatively as a framework for punishing breaches but positively in assisting professionals in recognising their own moral parameters. Clearly corrupt behaviour is subject to more than just policing by professional institutions and in some cases can be deemed criminal offences.

Where misconduct and unprofessional behaviours have been identified in the past, this has had negative implications on the way the construction industry has been viewed and, in some cases, this has led to public concern and government attention. To address this, there has been a concerted effort to improve confidence and trust amongst the general public in light of some cases of professional misconduct. Professional ethics, reinforced through professional standards of professional bodies and codes of practice, have played an important part in gaining this confidence and trust alongside respectability and integrity. However, these codes and standards need to be embodied in management practices and roles and responsibilities assigned to ensure compliance across ethics programmes.

Codes of conduct are sometimes anchored around the Seven Nolan Principles and embedded as core requirements for organisational financial regulations, and these are selflessness, integrity, objectivity, accountability, openness, honesty and leadership. Codes of ethics should reflect the practices and cultures which construction clients want to encourage for their respective organisations and project teams. The Code of Ethics Checklist devised by the Chartered Management Institute provides clear guidance for employees on what is expected of them in terms of ethical behaviors and practices. This sends a clear signal to other parties including customers and suppliers that unethical practices are not acceptable. To support their Codes of Conduct and Professional and Ethical Standards, the RICS have quite helpfully created a decision tree for proceeding or not proceeding with certain courses of action, and for decision making in these situations.

12.2.10 Implications in Practice for Ethics in the Construction Industry

There are many implications for the construction industry from unethical practices, which will include reputational damage and image considerations for organisations. Report of high-profile cases that have received widespread press coverage and highlighted cartels, collusion on tenders, blacklisting of subcontractors and other illegal practices which have led to widescale public condemnation of the construction industry. For this reason, when unethical practices and behaviour come to light and especially when they come to light in the public domain, this can be extremely damaging not just for the individuals and organisation involved but for the reputation of the industry as a whole. Notwithstanding the fact that such practices may only represent a small minority of case, there is the perception that such behaviours are commonplace in the sector. It is clear that unethical practice can take a negative toll on the performance and ultimately the outcome of construction projects. Negative consequences associated with poor outcomes are numerous and lead to assertions that the industry is inefficient, unsafe, wasteful, compromised on quality and cannot deliver to time and budgetary constraints. Quality and ethics are inextricably linked through a common care premise to do the right things and therein a proven way to improve competitiveness, reduce costs and increase client satisfaction. From case studies in Nigeria, this chapter has highlighted many examples of unethical conduct associated with failures during bidding processes, with reports of lowest price awards, at the expense of quality being commonplace. Other more serious examples of bribery and corruption have also been covered, especially in developing countries, which have served to undermine public confidence and hinder economic development and capital investment. For these reasons it has become an important priority for governments to address such practices to curb practices and cultures around unethical practices. This has put more centrally governed focus on creating policies designed to implement and police ethical reforms, and this could be through public procurement agencies. Notwithstanding this premise, low participation in training and education, certainly in developing countries, could possibly explain why the respondents felt that construction staff are not aware of the importance of ethics within their respective organisations. Accordingly, one of the contributory factors for unethical behaviours could be a lack of education and training of those individuals entering the industry. To respond to this deficiency, organisations should promote continued professional development in the workplace for the teaching and learning around ethics in practice to improve quality in construction outcomes and raise the reputation of the industry. In addition, it has been suggested that colleges and universities should tailor more academic modules and courses to the subject of professional ethics in an attempt to address this deficiency in knowledge and education. Furthermore, organisations should be encouraged to instigate continued professional development (CPD) of their employees to counteract unethical practices and behaviours within their companies. These organisations alongside public agencies and professional bodies, especially in developing countries, should also adopt strict rules and regulations to enforce the values of ethical standards and outlaw those deemed by be unethical.

12.3   Final Reflections, Overview and Closing Remarks

This book has discussed many different aspects and issues that influence and affect professional ethics and the repercussions that can arise from non-ethical practices, especially in the context of the UK construction industry. Construction clients, as leaders in the procurement of projects, should be leading the way in cultural changes to improve the reputation of the industry. In this pursuit, they should be aware of the importance of ethics, alternative definitions and interpretations of ethics, the reputation of the construction industry, codes of conduct and governance and regulations in avoiding bad practices. In conclusion, it would appear that measures to improve the practice of professional ethics, such as professional codes of conduct, have gone some way to improve the way the industry works but there are still far too many cases emerging of unethical practices that are blighting the sector. Although arguably these practices are emerging from a small minority of the sector, they are creating a bad press for the whole industry and further measures should be instigated by construction clients to address this dilemma. Traditional responses in the past, at an institutional level, have been based on governance, regulations and punishment for non-compliance and clearly these have had only limited success. Perhaps construction clients should be leading the way for a cultural change in the industry to train, educate and motivate construction individuals and organisations in what professional ethics entail, measures to ensure compliance and the benefits that they can bring for the sector. This could be achieved through more focus on further education and higher education course modules linked to professional ethics and continued professional development (CPD) through workshops and training events in the workplace. These measures will hopefully contribute to providing a more ethical environment for the industry to work within and reap great benefits not just for clients, all construction-related organisations and the building projects that they procure but for the future of the construction industry at large. It is accepted, however, that to bring about these cultural changes will take conviction, integrity and in some cases courage not to engage in established unethical practices. These improvements once ingrained within the industry could then reap massive rewards in providing a safer, honest, trusting and more enjoyable working environment for all.

Appendix A

Anti-Bribery Policy

1.1 Purpose

The purpose of this document is to define the Organisation Policy on anti-bribery. The policy is based on the Bribery Act 2010 ('Act'), which came into force on 1 July 2011, replacing and enhancing fragmented and complex existing laws which date from 1889, 1906 and 1916.

The United Kingdom signed an international treaty in 1998 to combat bribery and corruption and the new Act is intended to ensure that the United Kingdom is compliant with its obligations under that treaty.

The Act applies to offences committed on or after 1 July 2011 and the earlier statutes continue to apply to offences committed before 1 July 2011.

1.2 Scope

1.2.1 To Whom the Policy Applies

This Policy applies to all **Members of the Organisation, including subsidiary companies,** which means all employees and independent members of Council and its Committees, in particular:

- Members of the Organisation who pay or accept a bribe could be personally prosecuted.
- Senior managers are at risk of prosecution if they 'turn a blind eye' to bribes which are paid or accepted by the staff they supervise.

The Policy also applies to students where they are acting on behalf of the Organisation. Either in a paid or in a voluntary role.

It should be noted that 'Commercial organisations' can be prosecuted and, in addition to companies and partnerships, this expression is wide enough to include universities, schools and charities.

The Act does not only apply to individuals and commercial organisations in the United Kingdom but also to those who have a 'close connection' (business presence) with the United Kingdom.

Professional Ethics in Construction and Engineering, First Edition. Jason Challender.
© 2022 John Wiley & Sons Ltd. Published 2022 by John Wiley & Sons Ltd.

1.2.2 Definitions

1.2.2.1 What is a Bribery and Corruption?
1.2.2.2 Giving a Bribe or Active Bribery

Bribery refers to the offering or receiving an unearned reward to influence someone's behaviour. One example of bribery is a 'Kickback' – unearned reward following favourable treatment. **Corruption** is any unlawful or improper behaviour that seeks to gain an advantage through illegitimate means. Both are illegal.

- An offence is committed where an individual or the Organisation gives, promises or offers any financial or other advantage which is intended to induce or reward the improper performance of a public function or business activity or is made in knowledge or belief that acceptance of that financial or other advantage will itself amount to improper performance.
- It does not matter whether the financial or other advantage is given, promised or offered directly or through a third party.
- The financial or other advantage does not actually have to be given – the offer is sufficient to commit the offence.
- Reasonable and proportionate corporate hospitality should not be caught by the Act. However, entertaining which is disproportionate, lavish or beyond what would be reasonably necessary to 'cement good relations' may be evidence of intent to induce or reward improper performance.
- An offence is committed where an individual or the Organisation requests, accepts or agrees to receive a financial or other advantage which is intended to induce or reward the improper performance of a public function or business activity.
- It does not matter whether the financial or other advantage is requested, accepted or agreed directly or through a third party.
- The financial or other advantage does not actually have to be received – requesting or agreeing to accept it is sufficient to commit the offence.
- Reasonable and proportionate corporate hospitality should not be caught by the Act. However, entertaining which is disproportionate, lavish or beyond what would be reasonably necessary to 'cement good relations' may be evidence of intent to induce or reward improper performance.
- It does not matter who pays the bribe, e.g., if a director pays a bribe to ensure that his company is awarded a contract, that is still caught by the Act.
- An offence is committed where an individual or the Organisation gives, promises or offers any financial or other advantage to a foreign official which is intended to:

    i) influence that foreign official in the performance of his/her official functions; and
    ii) secure business or an advantage for the Organisation.

- It is only a bribe if the financial or other advantage is not required or permitted under local laws.
- A belief that local practice permits the payment is not a defence.
- There is no exception for facilitation payments (small bribes paid to facilitate routine Government action).

- The offence is committed where the Organisation fails to prevent any employee or other 'associated person' (see below) from committing active bribery or bribery of a foreign official on the Organisation's behalf.
- The bribe is caught even if it takes place outside the UK.
- The bribe must be intended to induce or reward the improper performance of a public function or business activity.
- The bribe must also be intended to secure business or other advantage for the Organisation.
- Knowledge of the bribe by the Organisation is irrelevant.
- The Organisation is automatically liable for the bribery unless it can show it had 'adequate procedures' (see below) to prevent the bribery.
- If the Organisation is a member of a company (e.g., joint venture or spin out) the Organisation is only liable for the corporate offence in respect of bribes paid by that company if that company is providing services to the Organisation and the bribes are paid for the benefit of the Organisation rather than the company itself.
- Someone who is providing the Organisation with Services so includes employees, agents, consultants, contractors and external partner organisations.
- The Act does not define 'adequate procedures'. Government guidance makes clear that each organisation needs to decide a proportionate response based on its activities and areas of risk (see Related Documentation).

1.3 Policy Statements

1.3.1 Position Statement and Commitment

Bribery is a criminal offence, is morally wrong and exposes the Members of the Organisation to the risk of prosecution, fines and imprisonment as well as endangering the Organisation's reputation.

The Organisation is committed to maintaining the highest ethical standards and to carrying on its activities fairly, honestly, openly and in compliance with all applicable laws.

Bribery will not be tolerated by the Organisation and all Members of the Organisation and Associates (which means [but not exclusively] agents, consultants, contractors, service providers, external partner organisations, suppliers, subsidiaries and joint venture partners) are required to uphold the highest standards of integrity in their dealings with or on behalf of the Organisation and to comply with all applicable laws of the countries in which they are working.

1.3.2 The Main Risk Areas for the Organisation

The main risk areas that have been identified for the Organisation are:

> *Purchases of goods and services by the Organisation, particularly the award of tenders and contracts*

(e.g., building projects).

Tenders and bids for research contracts which are made by the Organisation.

The giving/receiving of gifts and hospitality.

The admission/recruitment of students (particularly from overseas).

Collaborations, joint ventures, partnerships, affiliations (academic and commercial) (particularly with organisations based overseas).

Fundraising by the Organisation (typically where the donation of funds or sponsorship is conditional on the Organisation taking/not taking some action).

Student assessments/examinations.

Subsidiaries and spin-out companies.

Members of staff should not accept any gifts, rewards or hospitality (or have them given to members of their families) from any organisation or individual with whom they have contact in the course of their work that would cause them to reach a position whereby they might be, or might be deemed by others to have been, influenced in making a business decision as a consequence of accepting such hospitality. The frequency and scale of hospitality accepted should not be significantly greater than the Organisation would be likely to provide in return.

If any Members of the Organisation identify any bribery risk which they consider has not been addressed or adequately addressed by any actual or planned anti-bribery measures, then they should report this to the Director of Finance, either directly or through their line manager.

1.3.3 Anti-Bribery Measures

The Executive Committee is the principal body which approves the Organisation's anti-bribery measures, but these will also be reported to and endorsed by Council and, where applicable, Senate on an ongoing basis.

To promote a strong anti-bribery culture, the Organisation's existing policies and procedures in the main areas of risk have been reviewed and, where necessary, enhanced. The key policies and procedures for the prevention of bribery are listed in the Related Documentation section and include appropriate record keeping, internal controls as well as monitoring.

Appropriate anti-bribery measures have been embedded in all relevant operational policies and procedures across the Organisation. All relevant standard contracts and documents have also been updated to include anti-bribery provisions. It is particularly important for Members of the Organisation to:

1) Get to know every organisation which the Organisation is intending to work with – whether the organisation will be a collaborative partner, supplier, contractor, consultant or something else – and check that it has an embedded anti-bribery culture and has adopted similarly robust anti-bribery policies and procedures.
2) Use the up-to-date and applicable standard contract or document which incorporates anti-bribery provisions and do not amend any of those anti-bribery provisions without obtaining legal advice from the Organisation Solicitor.
3) Register gifts above £25 and hospitality above £100 (these limits apply to both single gifts and to the cumulative total of three gifts or more received within a rolling period of 12 months from the same source). All gifts and hospitality should be recorded by VCET members and Deans of Schools.

A training programme and communications strategy has been implemented to raise awareness of the Act and this Policy (see Related Documentation) and all new staff have to complete an online training session on the Bribery Act as part of their induction.

1.3.4 Offences under the Act

Three offences (see Definitions section above) apply to all Members of the Organisation:

> Giving a bribe (active bribery)
> Receiving a bribe (passive bribery)
> Bribing a foreign official

A further offence the 'corporate offence' (see Definitions) applies only to the Organisation.

Further guidance on the anti-bribery offences is included in the Appendices.

1.3.5 Reporting Bribery

All Members of the Organisation have a responsibility to help detect, prevent and report not only bribery but all other suspicious activity or wrongdoing.

All concerns or suspicions of bribery, fraud or corruption must be reported immediately so that the matter can be investigated, and appropriate action taken in accordance with the Counter Fraud Policy and Response Plan (see Related Documentation). In the first instance any concerns about bribery **must be reported to the Director of Finance**. If the issue relates to the Director of Finance, the matter must be reported immediately to the Chief Operating Officer.

1.4 Policy Enforcement

Any breach of this Policy will be regarded as a serious matter and is likely to result in disciplinary action or removal from office. Managers may also be subject to disciplinary sanctions for supervisory failures. The Organisation will also determine whether a breach of this Policy is such as to be referred to the Police.

The Organisation will avoid doing business with agents, consultants, contractors and suppliers who commit bribery and may end contracts with them.

Range of Possible Penalties (Criminal and Internal Sanctions)
For individuals:

1) A criminal record with a sentence of imprisonment of up to 10 years and unlimited fines.
2) Employees dismissal or other disciplinary sanctions.
3) Employees removal from office.
4) Damage to reputation.

For the Organisation:

5) A criminal record with unlimited fines and a potential ban from bidding for future research and other public contracts.
6) Damage to reputation and loss of public trust and confidence.

7) Adverse impact on donors, recruitment of Members of the Organisation.
8) Regulatory and funding issues.
9) Disruption to business activities.

1.5 Related Documentation

1.5.1 Related Policies

The following documents can be found on the Organisation Policy and Procedure pages:

- Financial Regulations (Financial Regulations and Accounts Section)
- Counter Fraud Policy and Response Plan (Financial Regulations and Accounts section) Human resources Section

  - Welcoming New Colleagues
  - Disciplinary Policy
  - Whistle-Blowing Policy

- Gift Acceptance and Ethical Fundraising policy (under Financial Management)
- Register of Interests, Gifts and Hospitality Policy (Declaration of management of Conflicts of Interest) (under Core and Organisational Governance)
- Code of Conduct for employees and executive

1.6 Training

Anti-bribery E-Learning toolkit within new staff induction section of Organisation website.

Appendix B

Counter Fraud and Response Policy

2.1 Scope and Introduction

The Organisation is committed to ensuring that high legal, ethical and moral standards are in place across the organisation, including where third parties act on the Organisation's behalf, at both home and abroad and is committed to countering any form of fraud or corruption.

It is therefore vital that measures are in place to ensure that there is an anti-fraud culture in the Organisation in which fraud is deterred, prevented and detected and that all suspected frauds are appropriately investigated and the necessary sanctions are imposed where a fraud is proven.

Organisation to have a robust and comprehensive system of risk management, control and corporate governance and that this should include the prevention and detection of corruption, fraud, bribery, money laundering and other irregularities.

This Policy sets out the roles and responsibilities of staff, Committees and other parties towards achieving this. Specifically, the sections which follow outline responsibilities for preventing and detecting fraud and set out how staff should respond if they suspect that a fraud is or has been taking place.

This Policy applies to all members of the Organisation including subsidiary companies and associated persons. Organisation includes employees and independent members of Council and its Committees.

The policy applies to any fraud, or suspected fraud, involving employees as well as consultants and contractors.

2.2 Definitions

Fraud – Fraud is legally defined within the Fraud Act 2007. For practical purposes, fraud may be defined as:

'The use of deception with the intention of obtaining an advantage, avoiding an obligation or causing loss to another party.'

Examples of Organisation fraud include:

- Misappropriation or theft of cash, stock or other assets – this might include the theft of stationery for private use, or the unauthorised use of Organisation vehicles, computers or other equipment.

Professional Ethics in Construction and Engineering, First Edition. Jason Challender.
© 2022 John Wiley & Sons Ltd. Published 2022 by John Wiley & Sons Ltd.

- Purchasing fraud – this can include approving or paying for goods not received, paying inflated prices for goods and services or accepting any bribe.
- Misstating claims or eligibility for other benefits – such as overstating or making false travel and subsistence claims.
- Accepting pay for time not worked – this can include failing to work full contracted hours, making false overtime claims, completing private work during Organisation time or falsifying sickness.
- Record fraud, often via computers – such as altering or substituting records, duplicating or creating spurious records or destroying or supressing them.
- Intellectual Property (IP) theft – such as claiming Organisation intellectual property as your own, or otherwise using or selling Organisation IP for your own personal gain.
- Academic fraud including immigration, admissions, internships, examinations and awards.

This list is illustrative and not exhaustive; other examples of fraud also exist. **Corruption** – dishonest or fraudulent conduct, typically involving bribery.

Bribery – the offering, giving, receiving or soliciting of any item of value to influence the actions of an official or other person in charge of a public or legal duty.

2.3 Policy Statements

2.4 Counter-Fraud Policy Objectives

The eight key objectives of the Organisation's counter fraud policy are:

1) Establishment of a counter-fraud culture.
2) Maximum deterrence of fraud.
3) Active and successful prevention of any fraud that cannot be deterred.
4) Rapid detection of a fraud that cannot be prevented.
5) Professional investigation of any detected fraud Effective internal and external actions and sanctions against people found to be committing fraud, including legal action for criminal offences.
6) Effective communication and learning in relations to fraud.
7) Effective methods of seeking redress when/where fraud has been perpetrated.

2.4.1 Fraud Prevention

8) The Organisation recognises the importance of prevention in its approach to fraud and has in place various measures including denial of opportunity, effective leadership, auditing and employee screening.
9) Fraud is minimised through usefully designed and consistently operated management procedures which deny opportunities for fraud. In particular, financial systems and procedures take into account the need for internal checks and internal control. Additionally, the possible misuse of information technology is prevented through the management of physical access to terminals and protecting systems with electronic access restrictions where appropriate.

10) The Organisation's Audit and Risk Committee provides an independent and objective view of internal controls by overseeing Internal and External Audit Services, reviewing reports and systems and procedures and ensuring compliance with the Organisation's Financial Regulations and the requirements of the OFS. These external reviews of financial checks and balances and validation testing provide a further deterrent to fraud and advice about system development/good practice.

11) The Organisation has in place a number of policies and related guidance that assist in preventing fraud. Please see Related Documentation.

2.5 Fraud Detection

Whilst it is accepted that no systems of preventative measures can guarantee that frauds will not occur, the Organisation has in place detection measures to highlight irregular transactions.

1) All internal management systems are designed with detective checks and balances in mind and this approach is applied consistently utilising wherever possible the expertise and advice of the Organisation's Auditors.

2) The approach includes the need for segregation of duties, reconciliation procedures, the random checking of transactions and the review of management accounting information including exception reports.

3) As set out in the whistle-blowing policy, concerns expressed by staff, or others associated with the Organisation are investigated by the Organisation without adverse consequences for the complainant, maintaining confidentiality wherever possible.

4) The Organisation views its preventative measures by management, coupled with sound detection checks and balances as its first line of defence against fraud.

2.5.1 Roles and Responsibilities for Preventing and Detecting Fraud

All Organisation senior managers and employees have a clear responsibility for the prevention and detection of fraud. The key responsibilities of individuals and groups are set out below.

A) Organisation Council and Audit and Risk Committee

1) The Council is ultimately responsible for ensuring that systems are in place for the prevention, detection and investigation of fraud, whilst day-to-day operation of relevant policies, procedures and controls is delegated to management.

2) The Council, together with the Audit and Risk Committee, is responsible for:

Adopting and approving a formal fraud policy and response plan.
Setting the framework with regard to ethos, ethics and integrity.
Ensuring that an adequate and effective control environment is in place.
Ensuring that adequate audit arrangements are in place to investigate suspected fraud.

B) Line Managers

1) Line managers are responsible for implementing this Policy in respect of fraud prevention and detection and in responding to incidents of fraud. In particular, this involves ensuring that the high legal, ethical and moral standards are adhered to in their School or Professional Service area. The practical requirements of line managers are to:

- Have an understanding of the fraud risks in their areas and to consider whether processes under their control might be at risk.
- Have adequate processes and controls in place to prevent, deter and detect fraud.
- Be diligent in their responsibilities as managers, particularly in exercising their authority in authorising transactions [electronically or otherwise] such as timesheets, expense claims, purchase orders, returns and contracts.
- Deal effectively with issues raised by staff including taking appropriate action to deal with reported or suspected fraudulent activity.
- Report suspected frauds according to the process outlined in Section 2.6.
- Provide support/resource as required to fraud investigations.

C) All Employees

1. The Organisation expects all employees to be responsible for:

- Upholding the high legal, ethical and moral standards that are expected of all individuals connected to the Organisation.
- Adhering to the policies and procedures of the Organisation.
- Safeguarding the Organisation's assets.
- Alerting management and/or other contacts should they suspect that the possibility of a fraud exists.
- Being aware of the Organisation policies and procedures to the extent they are applicable to their role.

D) Internal Audit

1) The Organisation's Internal Auditors are not responsible for detecting fraud. As with all aspects of governance, control and risk management is the responsibility of management.
2) However, Internal Audit's role in respect of fraud is to:

Regularly review fraud policies, procedures, prevention controls and detection processes making recommendations to improve these processes as required.

Discuss with management any areas which it suspects may be exposed to fraud risk.

Help determine the appropriate response to a suspected fraud and to support any investigation that takes place.

Facilitate corporate learning on fraud, fraud prevention and the indicators of fraud.

E) External Audit

External Audit is not responsible for detecting fraud. However, should the impact of fraud, as with all material misstatements, be of such magnitude as to materially distort the truth and fairness of the financial statements, the external auditors should detect the fraud and report it to the Audit and Risk Committee.

2.6 Response to Suspected Frauds

1) Members of staff are key to ensuring that the Organisation's stance on fraud is effective. All staff are positively encouraged to raise any concerns that they may have. All such concerns will be treated in confidence, wherever possible, and will be impartially investigated.
2) The information below sets out the detailed approach to reporting suspected frauds and how they will be investigated through to action and formal reporting. Please see Appendix 3 for a summary flowchart of this detail which covers all cases except those involving allegations against the Executive Director of Finance and/or Vice-Chancellor which is covered in Appendices 4 and 5.

2.7 Initial Report

1) If a member of staff believes that they have reason to suspect a colleague, contractor or other person of fraud or they are being encouraged to take part in fraudulent activity, they must immediately report this to their Line Manager.
2) If it is believed that this post holder is involved or an alternative reporting route is preferred, the Director of Finance should be informed. If the report comes via this latter route then the best approach to the investigation, considering the principles outlined below, will need to be considered and the Executive Director of Finance will liaise with other Senior Managers as appropriate.
3) Employees or managers should not initiate their own investigations or enquiries but should seek the advice of either the Executive Director of Finance or the Organisation Secretary as soon as possible. Appendices 1 and 2 provide some at-a-glance guide for employees and managers as to their role in responding to fraud.
4) If a member of staff believes that they have reason to suspect the Executive Director of Finance, they must immediately report this to the Chief Executive.
5) If a member of staff believes that they have reason to suspect both the Executive Director of Finance and Chief Executive they must report this to either the Chair of the Audit and Risk Committee.

2.8 Initial Investigation

1) The Executive Director of Finance and the Organisation Secretary will meet to consider the most appropriate response. This meeting should usually take place within 24 hours of the incident being reported.
2) Usually, an initial confidential investigation will take place with an appropriate investigating officer being appointed. Depending on the nature of the suspected fraud and the facts that have already been established, the Executive Director of Finance and the Organisation Secretary will consider reporting the suspected fraud to the police, internal audit, the Audit and Risk Committee Chair or others ahead of the initial investigation.

3) The purpose of the initial investigation is to gather all relevant information and documentation in order to determine if there is a prima facie case for further formal internal/external investigation. This investigation will be undertaken urgently and confidentially with a report being made to the Director of Finance and the Organisation Secretary.

4) The Executive Director of Finance and the Organisation Secretary will then consider whether:

1) There is a case for further investigation/action. If there is no case for further investigation/action, there should be an appropriate communication to the staff member who reported the suspected fraud.

2) There are immediate measures that would prevent any further losses including the suspension of staff.

3) Where appropriate, to approach external parties such as the internal or external auditors or specialist legal advisors, for advice on how an investigation of this type will proceed and to take advice on searching for, securing and preserving information, including documentary and electronic evidence and systems of all types.

4) To determine whether or not specialist expert advice will need to be engaged.

5) The matters reported constitute minor misconduct or other matters, which may be delegated for further investigation or management to other suitable managers using the appropriate Organisation policies and procedures. If this course of action is taken, the Director of Finance and the Organisation Secretary will retain overall oversight and may choose to take further formal action as evidence emerges.

6) In the case of allegations against the Executive Director of Finance the Chief Executive and Organisation Secretary will meet to consider the most appropriate response. This meeting should usually take place within 24 hours of the incident being reported. The initial investigation will then cover the same points as detailed in paragraphs 2.5.2 (b) to (d).

7) In the case of allegations against both the Executive Director of Finance and Chief Executive the Chair of Audit and Risk Committee, Chair of Council will meet to consider the appropriate response. This meeting should usually take place within 24 hours of the allegation being reported.

2.9 Further Formal Investigation

1) Where there is a case for further formal action or investigation, the Executive Director of Finance and/or Organisation Secretary will, as soon as reasonably practical, take steps to initiate a Formal Investigation. The Chief Executive and, if involving a member of staff, the Director of Human Resources, should be informed that this investigation is being carried out and should be kept appraised of its progress.

2) Under these circumstances, an individual, or group of individuals, should normally be advised of the concerns relating to them. Where those under suspicion are members of staff, the Disciplinary Policy should be adhered to.

3) At such time as an individual or group of individuals are advised of suspicions or allegations they will immediately be suspended and all access to internal files and papers (electronic and otherwise) will be disabled. Any prearranged meetings or tasks including planned visits to external locations should be reassigned to other staff. The Investigating Officer should seek advice on any such actions from the Director of Human Resources.

4) The Investigating Officer involved in the initial review shall, under normal circumstances, be requested to lead the formal investigation. The Executive Director of Finance/Organisation Secretary may consider appointing an external person to lead this work if it is more appropriate.

5) The Investigating Officer shall be provided with all assistance that he or she reasonably requires or requests including assistance with fulfilling their day-to-day duties which will be subordinated to the investigation.

6) The Investigating Officer may delegate tasks to other members of staff subject to ensuring that such members of staff maintain the confidentiality of the tasks assigned to them and, with the prior agreement of the Executive Director of Finance/Organisation Secretary wherever this prior agreement is practical to obtain.

7) The Investigating Officer will also consider whether external specialists are required to assist with the investigation such as forensic accountants or internal audit.

8) The Investigating Officer, as advised by the Organisation's Director of Human Resources or where appropriate based on legal advice, may communicate with appropriate members of staff for the purposes of gathering information and evidence and will, unless it will compromise the investigation, consult relevant senior staff of the School/Professional Service whose area the issue under investigation has arisen, always ensuring the maintenance of confidentiality.

9) The Investigating Officer shall liaise with and take advice from the Director of Human Resources over all matters related to the rights of staff potentially affected by the investigation including the alleged perpetrator. They will also aim to minimise disruption to operational activities and routines.

2.10   Formal Investigation Report

1) A formal report of the investigation and key outcomes will be presented to the Executive Director of Finance/Organisation Secretary as a basis for their decision upon any subsequent actions including any formal Disciplinary Hearing.

2) Liaison with the Police and potential legal action.

3) The nature and timeline of any system review.

4) Liaison with the Chair of the Audit and Risk Committee and the requirement to formally notify OFS.

5) Any suspension of an individual suspected of fraud will be carried out in accordance with the Disciplinary Policy. If a case of fraud is proven, the Organisation will act accordingly, and disciplinary proceedings may lead to dismissal.

6) The Organisation will seek prosecution of any individual where a criminal offence has been committed and the evidence obtained is sufficient to achieve a criminal

conviction. In addition, the Organisation will follow civil proceedings to recover money where appropriate.

2.11   Formal Reporting of Frauds

1) Any fraudulent activity will be reported on the fraud register irrespective of whether the Organisation suffered a financial loss.
2) The Organisation must report, without delay, any significant fraud (defined as those where the financial loss is over £25,000) or impropriety, to all of the following:

> The chair of the Audit and Risk Committee
> The chair of the Organisation Council
> The Internal Auditors
> The External Auditors
> OFS as a 'Reportable event'

3) The timing of such a report will depend upon the nature of the fraud and investigation. In all relevant cases, the Formal Investigation report should be summarised and provided to these individuals and bodies.

1) Audit and Risk Committee
It may be appropriate, subject to agreement with the Chair of the Audit and Risk Committee, to keep the Audit and Risk Committee itself appraised of an ongoing fraud investigation.

If this is the case, on completion of any Formal Investigation, a written report will be submitted to the Audit and Risk Committee containing:

a description of the incident, including the value of any loss, the people involved and the means of perpetrating the fraud the action that has been taken against the perpetrator(s) and the measures taken to prevent a recurrence; and, any action needed to strengthen future responses to fraud, with a follow-up report on whether or not the actions have been taken.

This report will normally be prepared by the Investigating Officer with external assistance where appropriate.

2) The Police
Consideration of whether and when to report an incident to the police will be taken by the Executive Director of Finance/Organisation Secretary and a report may be made at any stage during the investigation process.

Whilst reporting to the police of fraud or serious financial irregularity is likely to be the norm, depending on the nature of the incident, immediate reporting may not be appropriate until a body of material can be put before the police. It should be noted that under some types of insurance, a report to the police may be obligatory and this should be confirmed with finance.

The Investigating Officer shall liaise and co-operate with the police in any case where there has been a report to the police which the police decide to investigate.

All police contact, including the arrangement of visits by the police, shall be arranged through one of the Investigating Officer/Executive Director of Finance/Organisation Secretary unless otherwise delegated by them. Where the police ask to see members of staff or their work or records, the Director of Human Resources must first be involved before any visit is voluntarily agreed or arranged.

Where an information provider has approached the police directly, rather than the Organisation, with the report of a suspicion and the police contact the Organisation for further information, the enquiries should be referred to the Executive Director of Finance/Organisation Secretary before any further action is taken.

2.12 Managing Public Relations

Any requests for information from the press or anyone outside the Organisation concerning any investigation of irregularity must be referred directly to the Organisation Secretary. The advice of the External Communications team will be taken into consideration by the Organisation Secretary prior to issuing any statements. Under no circumstances should the Investigating Officer or other manager/employee provide statements to press or external persons.

2.13 Related Documentation

Finance Section:
  Financial Regulations [Financial Regulations and Accounts section]
  Anti-Bribery Policy (and guidance) [Financial Regulations and accounts section]
- Money Laundering Policy
  Criminal Finance Act Policy
  The following staff policies can be found on the HR Policies and Forms Pages:
  Whistle-Blowing Policy
  Disciplinary policy
  Register of Interests, Gifts and Hospitality Policy (Declaration and Management of Conflicts of interest)

Appendices

  Appendix 1: Guidance for Staff
  Appendix 2: Guidance for Line Managers
  Appendix 3–5: Flowcharts of Fraud Reporting and Investigation

Appendix 1: Guidance for Staff

Q. What should you do if you suspect a fraud?

- **Do make an immediate note of your concerns.** *Make a note of all relevant details, such as what was said in telephone or other conversations, the date, time and the names of any parties involved.*
- **Do convey your suspicions to someone with the appropriate authority and experience, commencing with your line manager. If this does not lead to a satisfactory response, then consider escalating the concern.** *Tell the Executive*

Director of Finance. *If it is believed that this post holder is involved or an alternative reporting route is preferred, the Organisation Secretary should be alerted.*

- **Do deal with the matter promptly.** *Any delay could cost the Organisation money or reputational damage. If in doubt, report your suspicions anyway.*
- **Do not be afraid of raising your concerns.** *Your concerns will be dealt with in confidence. You will not be ridiculed and will not suffer any recriminations as a result of voicing a reasonably held suspicion. The Organisation will treat any matter you raise sensitively and confidentially. We will ensure you receive appropriate support.*
- **Do not confront an individual or individuals with your suspicions and don't accuse any individuals directly.**
- **Do not try to investigate the matter yourself.** *There are special rules surrounding the gathering of evidence for use in criminal cases. Any attempt to gather evidence by people who are unfamiliar with these rules may compromise the case.*
- **Do not tell anyone about your suspicions other than those with the proper authority.** *All reported frauds will be investigated and if appropriate the police may be involved.*

Appendix 2: Guidance for Line Managers

- **Do be responsive to staff concerns.**

The Organisation needs to encourage staff to voice any reasonably held suspicions as part of developing

an anti-fraud culture. As a manager you should treat all staff concerns seriously and sensitively.

- **Do note details.**

Note all relevant details. Get as much information as possible from the reporting member of staff. If the staff member has made any notes, obtain these also. In addition, note any documentary evidence that may exist to support the allegations made. But DO NOT interfere with this evidence in any way.

- **Do advise the appropriate person according to the Whistle-blowing policy available on the staff channel.**
- **Do deal with the matter promptly.**

Any delay may cause the Organisation to suffer further financial loss or reputational damage.

- **Do not ridicule suspicions raised by staff.**

The Organisation cannot operate effective anti-fraud and whistle-blowing policies if staff are reluctant to pass on their concerns to management.

- **Do not approach or accuse any individuals directly.**
- **Do not convey your suspicions to anyone other than those with the proper authority.**
- **Do not try to investigate the matter yourself.**

Remember that poorly managed investigations by staff who are unfamiliar with evidential requirements are highly likely to jeopardise a successful criminal prosecution.

Appendix C

Criminal Finances Act Policy

3.1 Scope and Introduction

The Organisation is committed to ensuring that high legal, ethical and moral standards are in place across the organisation.

The Criminal Finances Act (CFA) 2017 came into force on 30th September 2017. Part 3 of the Act introduces a new 'corporate criminal offence of failure to prevent the facilitation of tax evasion'. The legislation applies to all business and all taxes. This particular offence is not about the Organisation itself avoiding, evading or underpaying tax, *but about the Organisation failing to prevent its employees/ agents/ associates from facilitating the evasion of tax by another party.*

All UK corporates are affected and can be subject to prosecution for the facilitation of tax evasion by 'associated persons'.

There are two corporate offences – a domestic tax fraud offence and an overseas fraud offence. A UK incorporated body can be prosecuted for either offence. While the overseas offence is slightly narrower in scope, it still essentially means that universities need to consider the potential for overseas tax evasion as well as UK tax evasion.

The UK Offence:
This requires three stages:

1) The criminal tax evasion by a taxpayer under existing law.
2) The criminal facilitation of the tax evasion by 'an associated person' of the relevant body who is acting in that capacity (as defined by the Accessories and Abettors Act 1861).
3) The relevant body failed to prevent its representative from committing the criminal facilities action (legislation from the 30th September 2017).

The Overseas Offence:
This requires the same three stages, but there are additional questions to consider determining whether it is an offence under CFA:

1) The criminal tax evasion by a taxpayer (either an individual or a legal entity) under existing law The criminal facilitation of the tax evasion by an 'associated person' of the relevant body who is acting in the capacity (as defined by the Accessories and Abettors Act 1861)

Professional Ethics in Construction and Engineering, First Edition. Jason Challender.
© 2022 John Wiley & Sons Ltd. Published 2022 by John Wiley & Sons Ltd.

1) Stage a – Would this be a crime if carried out in the UK?
2) Stage b – Does the overseas jurisdiction have the equivalent laws at Stages 1 and 2?

2) The relevant body failed to prevent its representative from committing the criminal facilities act.

A successful prosecution could lead to:

- An unlimited fine
- A public record of conviction
- Significant reputation damage and adverse publicity

There is a defence of having reasonable prevention procedures in place.

This Policy applies to all members of the Organisation Community including subsidiary companies and associated persons. Organisation community includes employees and independent members of Council and its Committees.

Examples of situations where Organisation employees, agents and associates could be considered to be assisting third parties to evade tax can be found in Appendix 1 to this document.

3.2 Definitions

A Relevant Body – an incorporated body or partnership

An Associated Person is an employee, agent or other person who performs services for or on behalf of the relevant body. The offence is committed where the facilitation offences are committed by someone acting in the capacity of an associated person. The associated person can be an individual or incorporated body.

3.3 Organisation Commitment

The Executive Team has endorsed the following statement:

'At the Organisation, we do not condone and have a zero-tolerance approach to the facilitation of tax evasion. Tax evasion occurs where employees, agents or businesses providing services for or on behalf of the Organisation omit, conceal or misrepresent information to reduce their tax liabilities.

As part of our commitment to enforcing Criminal Finances Act 2017 the Organisation, including subsidiary companies, will maintain reasonable and proportionate processes and procedures to prevent fraudulent activity by its staff and anyone acting on its behalf from criminally facilitating tax evasion in the UK and/or overseas.'

All employees of the Organisation and its subsidiaries have a duty and responsibility to support this approach.

3.4 Policy Statements

1) The Organisation has nominated key officers responsible for the information and queries on CFA within our organisation – the key officers for CFA 2017 are the Executive Director of Finance and Head of Financial Accounting.

2) The Organisation's Head of Procurement is responsible for supply chain transparency and initiatives.
3) Training and awareness sessions will be undertaken for all Finance staff and where the Organisation identifies CFA2017 specific risks it will undertake bespoke training and awareness sessions.
4) The Organisation regularly reviews its risks and associated processes and procedures to ensure all steps are taken to prevent facilitation of tax evasion.
5) The Organisation regularly reviews guidance and legislation (at least twice a year) in relations to CFA 2017 to ensure it is maintaining an appropriate CFA 2017 policy.
6) The Organisation maintains a register of possible risks of the facilitation of tax evasion by its staff and associates (including agents, contractors, suppliers and intermediaries), as well as listing controls to mitigate those risks, and any actions required to improve these controls. The register is regularly reviewed and updated, as and when required in relation to the nature of the specific risks.
7) Appropriate due diligence will be undertaken on both customers of and suppliers to the Organisation and its subsidiary companies. This due diligence will be proportionate to the level of perceived risk of the interaction being used to engage in the criminal facilitation of tax evasion.

3.5 Roles and Responsibilities

3.5.1 Organisation Council

The Council are ultimately responsible for:

- Approving the policy that Organisation does not condone and has a zero-tolerance approach to the facilitation of tax evasion.
- Adopting and approving a formal policy and response plan for alleged breaches.

On the recommendation of the Audit and Risk Committee, Council are responsible for:

- Ensuring that an adequate and effective control environment is in place.
- Ensuring that adequate audit arrangements are in place to investigate suspected concerns.

3.5.2 Line Managers

Line managers are responsible for implementing this Policy. In particular, this involves ensuring that the zero-tolerance approach to the facilitation of tax evasion is adhered to in their School or Professional Service area. The practical requirements of line managers are to:

- Have an understanding of the potential risks in their areas and to consider whether processes under their control might be at risk.
- Ensuring that agents and associated persons are aware of their responsibilities under the CFA.
- Have adequate processes and controls in place to prevent, deter and detect breaches of policy.

- Be diligent in their responsibilities as managers.
- Deal effectively with issues and concerns raised by staff including taking appropriate action to deal with reported or suspected breaches.
- Report suspected breaches according to Counter Fraud Policy and Response Plan.
- Provide support/resource as required to investigations.

3.5.3 All Employees

The Organisation expects all employees to be responsible for:

- Adhering to the policies and procedures of the Organisation including having a zero-tolerance approach to the facilitation of tax evasion.
- Alerting management and/or other contacts should they have concerns or suspect that the possibility of a breach exists.
- Being aware of the Organisation policies and procedures to the extent they are applicable to their role.

3.6   Response to a Suspected Facilitation of Tax Evasion

Members of staff are key to ensuring that the Organisation's stance on facilitation of tax evasion is effective. All staff are positively encouraged to raise any concerns that they may have. All such concerns will be treated in confidence, wherever possible, and will be impartially investigated.

3.7   Related Documentation

The following document can be found on the Organisation Policy and Procedure Pages (within relevant subject areas) accessible under 'P' from the Staff Channel A-Z:

- Financial Regulations
- Counter Fraud Policy and Response Plan

Appendix 1 – Examples of Possible Fraud/Evasion

The following are examples of situations when a Organisation employee or associated person would act in a manner to cause the Organisation to breach of the Criminal Finance Act.

1) Deliberate mis-categorisation of an individual as a self-employed contractor rather than employed.

When the Check Employment Status for tax (CEST) tool is completed it is important that this is completed according to the **true** nature of the relationship with the Organisation. Falsely and deliberately tweaking the answers on the tool so that a response of 'Self-employed' is achieved will result in HM Revenues and Customs (HMRC) collecting less National Insurance and potentially lower income tax.

2) Collusion with an overseas agent so that the payment is made to bank account which is not in the name of the agent or company or to a jurisdiction where the agent does not live or work.

Collusion could allow the overseas agent to reduce or avoid payment of tax. Supplier 'inconsistencies' need to be queried during the supplier set up process and internally escalated to the Head of Financial Accounting as appropriate.

3) Payment to one third party entity knowing that the goods/services have been provided by another entity with the primary purpose of evading tax.

Payments to third parties who have not provided the service, other than factoring arrangements, should be queried and internally escalated to the Head of Financial Accounting as appropriate.

4) Approving a VAT invoice for payment when it is known that the supplier is not VAT registered. If the supplier is not VAT registered, then the Organisation should not be paying VAT on invoices.

Collusion with a third party to deliberately mis-describe a supply of goods or services as a grant rather than a supply of goods or services.

5) This would result in VAT not being charged on the supply resulting in lower VAT recovery for HMRC. Care must be taken to ensure that grants are genuine grants and not supplies of goods and services – queries should be raised with the Research and Enterprise teams as appropriate.

Collusion with an overseas education establishment to misdescribe the services provided by the Organisation to the overseas establishment so that the overseas establishment avoids having to pay local withholding tax.

6) The overseas establishment advises an organisational member of staff that by misdescribing the 'teaching services' provided by the Organisation, the overseas establishment can avoid paying overseas withholding tax. This would result in a breach of the Criminal Finance Act as the overseas jurisdiction would recover lower tax than it is entitled to.

7) Falsely completing zero-rating certificates for medical research when the research is not of a medical nature.

The Organisation can obtain zero rating on the purchase of certain goods and drugs when undertaking grant-funded medical research. Deliberately miscategorising 'non-medical research' as 'medical research' on zero-rating certificates would result in a breach of the Criminal Finance Act as HMRC would not receive all the VAT that is due.

8) Colluding with an individual so that a payment for goods/services is described as a donation so that the donor can claim tax relief.

To be classed as a 'donation' the funds must have been freely given with no obligation on the Organisation to provide goods or services to the donor. Deliberate misdescribing the service will result in HMRC not recovering all the Income tax that is due.

9) Colluding with a donor so that when a donor offers a donation to the Organisation it is registered in their partner's name.

By registering a donation in someone other than the actual donor's name the partner may then claim gift-aid tax relief on the donation despite not being entitled to the relief. This will result in HMRC not recovering all the income tax that is due.

10) Entering into barter arrangements with third parties.

A Organisation member of staff arranges for an academic,not employed by Salford, to undertake work at the Organisation and in return offers the academic free use's facilities or to pay for attendance at conferences. Bartering arrangements need to be avoided as these result in HMRC not receiving income tax and VAT that may be due.

Appendix D

Template for the Gateway 1 Project Proposal

The writings in italics are for guidance only. Please delete them when you have completed your Mandate.

*This form is an initial summary of the reason(s) for the proposed project. It is required that sufficient information should be provided to enable decisions to be made on whether the proposal can proceed to the next stage of the Business Case approval process. The form is to be completed principally by the Project Manager in consultation with the Project Sponsor. A guidance document is also available and may be accessed **from this link.***

STRATEGIC-LEVEL DEVELOPMENT (i.e., initiated at institutional level)	YES	NO
	☒	☐

School/Service:

Project Title:

Programme: *If the project is a part of a programme*

Funded by: *Please indicate the main funding source for the project*

Project Sponsor:

1) STRATEGIC FIT/RATIONALE
Please explain the project in context by providing the reasons for wanting the project. Also include a brief statement of how the project fits into current client's strategy.

2) SCOPE AND OBJECTIVES
List what the project aims to achieve in terms of objectives and outline the scope of the project as project deliverables. Include anything which is excluded from the project (out-of-scope).

3) SUMMARY OF KEY BENEFITS
List the benefits that will be realised by achieving the scope of work. Ensure that you emphasise the key benefits, for instance by highlighting the income streams that will be grown, cost/efficiency savings, services that will be refined, etc. Include also how the benefits will be measured/demonstrated.

4) PROPOSED TIMESCALES
Provide a high-level timescales for project execution and implementation. Include the estimated time for completing the project proposal, business case and key milestones.

This is indicative only but should be based on what is currently believed to be achievable.

5) STRUCTURE OF GOVERNANCE

List the key stakeholders for the proposed project. Also indicate what consultation is needed or have already taken place.

6) PROJECT DEPENDENCIES

List any other project whose success depends on the successful delivery of the proposed project, or other externalities on which the proposed project depends on.

7) PROJECT RISKS

Outline the major risks which will need to be managed for the project.

8) PROJECT COSTS

Provide a high-level summary of the likely costs for the project to provide an idea of scale and affordability. Also include the cost type, e.g., fixed and variable. This is indicative only but should be based on what is currently believed to be achievable.

9) RESOURCES REQUIRED

List the key resources that are required to complete the project proposal stage of the business case approval. These include both financial and human resources and indicate whether internal or external capability will be utilised.

Appendix E

Template for the Gateway 3 Business Case Process Business Case

Business Case (Gateway 3) Template

Purpose of Document

The purpose of this document is to set out in detail the justification for the undertaking of a project based on the strategic objective, cost of the development and the identified benefits the Client will see post implementation. The business case should clearly say why the effort and time will be worth the investment. It should also provide reassurance that the project will be managed throughout the implementation stage and how the benefits identified will be measured. Provide reference to the approved Project Mandate and Project Proposal documents.

Project Details

Project Name
Project Sponsor
Project Manager
Proposed Start Date
Proposed Completion Date

Document Control

Version No.	Date	Details of Change	Authors

Professional Ethics in Construction and Engineering, First Edition. Jason Challender.
© 2022 John Wiley & Sons Ltd. Published 2022 by John Wiley & Sons Ltd.

Table of Contents

The writings in italics are for guidance only and should be deleted once you have completed your form. Further information is available in the associated Business Case guidance document.

Executive Summary

The executive summary is a high-level summary of the Business Case. It should include pertinent information to convey to the reader an understanding of the whole document in a clear and concise manner. It may be organised using the Business Case Section headings.

Section 1: Strategic Context: The Case for Change

This section describes the current situation and explains the need for the project from the perspective of the sponsoring Department, Service or School.

1.1   Organisational Overview

Please provide an overview of the sponsoring Department, Service or School. The overview should include Strategic Goals and Objectives, Current activities and Services, a high-level organisational structure and key stakeholders and clients. Existing capacity (financial and human resources may be included at discretion.

1.2   Project Background/Rationale

Please describe the background to the project as well explain the current state of affairs including the issues or opportunities which the project seeks to address (Business need). Include any changes since the submission of the Project Proposal (Gateway 2).

1.3   The Drivers for Project

Please identify both internal and external drivers for the project and link them to the business need.

1.4   Risk and Impacts of Not Continuing with the Project

List the consequences of not proceeding with the project.

Section 2: Strategic Context: the Solution

This section explains the nature and purpose of the project and how the project can solve the problems identified in Section 1 of the Business Case. It describes the project deliverables, benefits, alignment with the overall strategic objectives of the Client, costs, risks and other considerations which can affect the successful delivery of the project.

2.1 Project Purposes

The project purposes subsection describes what the project is all about and the expected outcome from the project.

2.1.1 Project Goal
Please provide a simple statement on what the project is about.

2.1.2 Project Objectives
Please break down the project Aim into specific objectives to enable the achievement of the stated Aim. Objectives should be SMART (specific, measurable, achievable, realistic, time-bound).

2.1.3 Project Outcomes
Please describe at a high level the expected project outcome(s). This is not the project output but rather the desired state of affairs, in other words, what the project is intended to achieve. The project outcomes provide a link between the project objectives and the benefit to be derived from the project.

2.2 Project Scope

Please describe the boundary (in-scope, and out-of-scope) of the project. The description should be in terms of identified business needs and any changes since submission of the Project Proposal. This may be considered along a continuum that may include the following:

Table 1 Project scope boundaries.

Project Boundary	Included	Excluded
Minimum scope: Essential requirement/project outcomes		
Intermediate scope: Essential and desirable project outcomes		
Maximum scope: Essential, desirable and optional project outcomes		

Consider if there are overlaps with other projects. Further information is provided in the Business Case Guidance document.

2.3 Critical Success Factors (CSFs)

Please list and quantify the CSFs for the project against which the successful delivery of the project will be assessed as well as used in the evaluation of options. Further information is available in the Business Case Guidance document.

2.4 Strategic Fit

Please demonstrate how the project fits within the broader Client strategic context and its contribution towards the achievement of Client priorities by completing the Table below.

Rating Legend

Blank	Low	Medium	High
Not applicable to the priority	Possible or uncertain contribution to the priority	Probable contribution to the priority	Clear and valuable contribution to the priority

Benefit cross reference

In this box identify which benefits A,B,C,D,E,F,G H I or J (Section 2.5) will impact upon which strategic objective.

Key measures

Measures to be adopted to measure that the project delivers the identified benefits. These can be existing KPIs, other key measures, customer satisfaction indices or a newly introduced measure specific to the project.

2.5 Project Benefits and Benefits Realisation Plan

2.5.1 Main Benefits Criteria (if Applicable)
Please describe in detail the main benefits associated with satisfying the scope of the project. Where possible this could be expressed in monetary form but bear in mind that some benefits may be qualitative in nature.

2.5.2 Main Project Dis-Benefits
Please describe any dis-benefit associated with the project.

2.5.3 Benefit Realisation Register
*Please complete the project benefits register as an appendix to this document. The register sets out the responsibility for the delivery of benefits, their measurements and time frame for realisation. The benefits register template may be downloaded **from here**.*

Table 2 Project strategic fit.

Sub-Strategic Priorities	Key Benefits and Measures	H	M	L
1:1 Academic Growth and Diversification: Increase in the number of students' enrolment in all areas: UG, PGT, PGR and Overseas.	Key benefits Key measures			
2:1 Education and Student Experience: Programme/courses development with industrial partnership	Key benefits Key measures			
2:2 Education and Student Experience: Provide opportunity for students to develop as active partners, become self-aware and responsible; provide opportunity for wider learning and contact to students both with peers and internal external facilitators.	Key benefits Key measures			
2:3 Education and Student Experience: Opportunities to develop students' graduate skills, attitudes, personal competencies and attributes	Key benefits Key measures			
3:1 Research and Enterprise: High-quality research with both nationally and international impact; increased research and commercial income	Key benefits Key measures			
3:2 Research and Enterprise: Development of new and existing institutional and industrial partnerships	Key benefits Key measures			
4:1 International Priorities: Increase in international students' enrolment	Key benefits Key measures			
4:2 International Priorities: Reputational risk management and enhancement of the Client's international profile	Key benefits Key measures			

2.6 Costs

The costs of the project should be estimated using the standard cost estimation practices. Please indicate the person responsible for estimating the project costs and the standard method used. While a summary of the estimated costs of the project is required to be provided in the main body of the Business case, a detailed breakdown of costs is to be also provided as an appendix to the document.

2.7 Timeline

Please provide an overview of the timescale for the project. Highlight deadline dates associated with key milestones.

2.8 Project Constraint

Please provide a summary of the constraints within which the project is expected to be undertaken. Also highlight any changes since the submission of the Project Proposal.

Table 3a Project benefits.

PROJECT BENEFITS

Type	Tick as Appropriate	✓	Please Explain and Include Value as Applicable	One-Time Value (£)	Annual Value (£)
Financial Benefits	A Cash releasing, e.g. cost avoidance				
	B Non-cash releasing, e.g. staff time saved				
	C New income				
	D Additional income				
Non-Financial Benefit	E Strategic fit		Cross reference with Section 2.4 to identify the strategic fit.		
	F Competitive advantage				
	G Competitive response				
	H Operational improvement/ management				
	I Others, e.g. staff moral				
Risk Avoidance	J			One-time value (£)	Annual value (£)

Additional Information

Table 3b Project dis-benefits.

Description	Stakeholder Group (as Applicable)	How and When
Dis-benefit 1		
Dis-benefit 2		
Dis-benefit 3		

Table 4 Summary of the estimated project costs.

	Year 1	Year 2	Year 3	Year 4	Year 5
Estimated Capital Costs (as Applicable)					
Hardware					
Software					
Others					
Total					
Estimated Revenue Costs (as Applicable)					
Staff costs – external					
Staff costs – internal (backfill)					
Maintenance charges					
Other non-pay costs, e.g. contingencies, inflation costs					
Total					

Please provide advice on the following:

1) Anticipated source of funding for the project
2) Impact on Client's borrowing/gearing
3) Any applicable funding deadlines
4) Possible cash savings
5) Possible additional income
6) The impact on staff full-time equivalents (FTEs) and student numbers
7) Procurement issues/decisions

Table 5 Summary of project constraints.

Project	Brief Description of Constraint
Constraint 1	
Constraint 2	
Constraint 3	

For each of the above, any changes since the submission of the Project Proposal should be highlighted.

2.9 Project Dependency

Please list all project dependencies (if applicable) which need to be managed for the entire duration of the project. Any changes since the submission of the Project Proposal should be highlighted.

2.10 Project Assumptions

Please list and describe all assumptions, for instance with scope definition, benefit realization and cost, and the potential impact they could have on the project if not addressed. The Table below could be used for the purpose

Table 6   Summary of project dependencies.

Project	Type of Dependency	Brief Summary of Linkages
Project A	Dependent on this project	Requires delivery of output x, y and z from this project
Project B	This project depends on B	Requires project B to deliver products a, b and c

Table 7   Summary of project assumptions.

Number	It Is Assumed that:	Effects on Project	Reliability Level: High/Medium/Low
Assumption 1			
Assumption 2			
Assumption 3			

N.B.: Assumptions, constraints and dependencies are significant sources of risks and issues for any project. It is therefore important that the relationship among them are consistently made clear. The information is both useful in the planning stage of the project and during options analysis. They also have a big impact on benefit realisation if proved to be false or unreliable.

2.11   Project Risks

Risk Identification
Please identify the potential risks to the project and assess them in terms of their probability of occurrence and impact. Consider risk for the entire life cycle of the project including both project delivery risks and project outcomes risks. More information is available in the Business Case guidance document.

2.11.2   Risk Assessment
Please assess the identified risk in terms of their impact and probability of occurrence. A risk assessment matrix (RAM) should also be produced to prioritise the identified risks. Based on your assessment, please categorise the assessed risks as minor, medium and/or major.
   You may consider producing the table below in a landscape layout.

Table 7a   Risk identification for the project.

Project Delivery Risk				
Risks ID	Risks	What Is at Risk	Source (How Can the Risk Occur)	Impact (What Would Be the Effect of the Risk)

Table 7b   Risk assessment for the project.

Risks ID	Probability (P)	Impact (I)	Risk Score (PXI)	Risk Category	Tolerance Rating*

*Each risk assessed as major or medium should be rated as one of the following: Acceptable, Unacceptable or Unknown. Please refer to the **Business Case Guidance document for** additional information.

2.11.3   Risk Register

Please complete the initial Risk Register as an appendix to this document. The Risk should describe the attributes of each major and medium risks as well as assign responsibility for the management of identified risks. The information in the risk register for each risk should include a risk ID, risk statement, impact/probability rating, risk prioritisation level, mitigation action, risk status and risk owner. A template is provided and can be downloaded from this link.

Section 3: Sustainability Impact Analysis

This section describes the impact the project will have on project stakeholders (including the Client) and on sustainability.

3.1   Sustainability Impact

Sustainability remains the key enabler for the Client Strategic plans. Please use this section to summarise the impact of the project on sustainability under the following sub-headings:

- **Financial Sustainability:**
- **Environmental Sustainability**
- **Social Sustainability**

3.2   Sustainability Action Plan

If applicable please state what actions are being put in place to remove/mitigate any potentially negative impact on sustainability.

Section 4: Options Appraisal Summary

This section assesses the various options that could address the identified business need including the preferred option. The business case is written on the basis that the preferred option is to be adopted.

4.1 Summary of Options Considered

The preferred option for the project is contained within the body of the Business Case. A summary of other options considered are provided below.

Please list and summarise all the options considered to meet the identified business need including the do-nothing option. The do-nothing option should describe what will happen if the project did not go ahead.

For each option identified, please carry out a SWOT analysis of potential options to narrow down your option to three possible options.

Options	Strengths	Weaknesses	Opportunities	Threats
Option 1				
Option 2				
Option 3				
Option 4				
Option 5				

Summary Option Ranking

Option 1				Possible
Option 2				Possible
Option 3				Possible
Option 4				**Discounted**
Option 5				**Discounted**

4.2 Option Evaluation Criteria

The Table below is designed to assist in the assessment of the possible options by considering the financial implications and the non-financial contributions of the three possible options identified above. Please complete it and rank your options based on performances on the indicators.

Financial Appraisal	Option 1	Option 2	Option 3
Financial Appraisal			
Total project cost			
Total additional on-going costs			
Quantifiable cash benefits			
Net Present Value (NPV)			
Non-Financial Appraisal (Critical Success Factors)			
Project outcome/benefit			
Contribution to client KPI			
Increase student numbers			

Financial Appraisal	Option 1	Option 2	Option 3
Improve NSS?			
Improve space utilisation			
Option Ranking	**3rd**	**2nd**	**1st**

4.2.1 Cost–Benefit Analysis

A further cost–benefit analysis may be conducted at discretion to further assist in the selection of a preferred option. More information is provided in the Business Case guidance document.

4.3 Rationale for Preferred Option

Please explain the reasons for choosing the preferred option as well as for discounting other options. Consider this in the light of the result of the evaluation criteria.

Section 5: Implementation Plan

This section describes the project management methodology and approach to be adopted for the delivery of the project.

5.1 Project Governance

Please summarise the key roles and responsibilities for those involved in the delivering of the project including the reporting arrangement.

Also advise whether a steering committee or Project Board will be required. If applicable please outline what representatives will be required.

5.2 Project Management Strategy

Please summarise the project management approach to be adopted to managing the project

Table 9 Summary of project governance.

Roles	Name of Personnel	Responsibility (What Each Have to Deliver to the Project)
Steering committee/project board		
Project sponsor		
Project manager		
Project team		
Project management Office		

5.3 Project Milestones Plan

Please describe the key things that need to be delivered in order to meet identified deadlines. More information is available in the Business Case guidance document.

5.4 Risk Management Plan

Please summarise risk management strategy to be adopted for the entire project duration. The risk register will be useful in this respect.

5.5 Change Control Strategy

Please describe what procedures are to be put in place to manage changes in project requirement otherwise known as scope creep.

5.6 Project Assurance Strategy

Please summarise the mechanisms to be put in place to monitor the performance of the project, for example the schedule of internal and external reviews.

Appendix

Please append supporting documents to the Business Case here.

1. Costs and resources template
2. Benefit register
3. Risk register
4. Option evaluation criteria analysis
5. Others

Project Reviewer Decision to Proceed

Please select one of the three options below.

Project name:
Date of decision   **Reference ID**

Option: Yes to proceed
Comments:

Proceed to full definition and Implementation   Yes/No   **Date**

Option: To review and resubmit proposal
Comments:

To be resubmitted to project Board on   **Date**

Option: No project declined
Comments:

Decision made by (Signatures)

Date:

Decision made by (Signatures)

Date:

Appendix F

Example of a Contractor Competency Questionnaire

6.1 Contractor Competency Questionnaire

As the client, the (organisation removed for confidentiality) has a duty under the Health and Safety at Work etc. Act 1974 (HASAWA) and the Construction (Design and Management) Regulations 2015 (CDM) to ensure that the Contractor engaged for work is competent to meet the obligations conferred on them by law.

To ensure we can meet the requirements of this duty the (organisation removed for confidentiality) use the Safety Schemes in Procurement Registered Members Scheme (SSIP) as a basis for evaluating contractor competence. If you do not have a current listing in SSIP that relates to the types of work required by the project, please complete this questionnaire and provide evidence in support of your submission.

Declaration	
I confirm that the information I have given in this form is a true and accurate statement of my organisation's Safety Management procedures.	
Name of person completing this form:	
Name of organisation:	
Position in organisation:	
Signature and date:	
Section 1 – COMPANY INFORMATION	
Name of Organisation:	No. of Employees:
Address	
Contact Name and Position:	
Tel. No.:	E-mail:
Insurance:	Insurer:
Employer's Liability Insurance held:	Policy No.:
Attach a copy of your policy	Extent of Cover:
Insurance:	Insurer:
Public Liability (3rd party) insurance held:	Policy No.:
Attach a copy of your policy	Extent of Cover:

Professional Ethics in Construction and Engineering, First Edition. Jason Challender.
© 2022 John Wiley & Sons Ltd. Published 2022 by John Wiley & Sons Ltd.

Section 2 – HEALTH and SAFETY INFORMATION

1   **Health and Safety Policy**

You are expected to have implemented an appropriate policy, regularly reviewed and signed off by the Managing Director or equivalent.

Attach a signed current copy of your company's health and safety policy (including statement of intent, organisation for health and safety and arrangements for H&S management within the organisation relevant to the nature and scale of the work).

2   **Competent Advice**

Your organisation and employees must have ready access to competent H&S advice. The advisor must be able to provide general, as well as construction related, H&S advice.

Please provide the contact details below:

Name:	
Address:	
Tel No.:	E-mail:

Provide details of their health and safety qualifications, experience and relevant training that enables them to undertake this responsibility:

3   **Risk assessment leading to safe methods of work**

You should have procedures for carrying out risk assessments and for developing and implementing safe systems of work/method statements.

Attach sample risk assessments and method statements undertaken for similar projects.

Section 3 – TRAINING and INFORMATION

4   **Training, Information and Instruction**

You should have training arrangements in place to ensure your employees have the skills and understanding necessary to discharge their duties as Contractors. You should also have a program of refresher training, to keep employees updated on new developments and changes in legislation to good H&S practice.

How does your Company provide relevant health and safety training, information and instruction to its employees?.

5   **Individual Qualifications and Experience**

Employees are expected to have the appropriate qualifications and experience for the assigned tasks, unless they are under controlled and competent supervision continuously.

Attach copies of qualifications or training certificates relevant to the assigned tasks, e.g. Asbestos Awareness, CSCS, ECS, CPCS, SMSTS, SSSTS, CSIRS, PASMA, IPAF, S/NVQs, and company-based training programme suitable for the work to be carried out.

6   Trades and Professional Bodies

Please provide details of any health and safety organisations, groups, associations or bodies that your company is a member of:

Please provide details of any trade federations or other professional industry bodies or organisations that your company is a member of:

Section 4 – HEALTH and SAFETY PERFORMANCE

7   Accident Reporting

You should have records of all RIDDOR reportable events for the last 3 years. You should also have a system for reviewing all incidents, and recording the action taken as a result.

Incidents reported to HSE under RDDOR					
Year	Fatalities	Specified Injuries	Over 7-Day Injuries	Dangerous Occurrences	Non-Reportable
This year (year to date)					
Last year (full calendar year)					
Year before last					
Enforcement action					

How do you investigate accidents and incidents?

8   Enforcement Action

You should record any enforcement action taken against your organisation over the last 5 years and the actions taken to remedy the relevant matters.

Provide details of any enforcement notices issued or prosecutions taken against your organisation in the last 5 years and what action was taken to rectify matters, if none please state:

Section 5 – MONITORING, AUDIT and REVIEW

9   **Monitoring of Procedures**

You should have a system in place for monitoring of procedures, for auditing them periodically and for reviewing them on an on-going basis.

What methods are in place to ensure effective communication/consultation with staff, on health and safety matters, following a procedural review?

How do you ensure health and safety procedures are followed by staff?

Attach an example of a site inspection report from a similar project.

Attach a copy of a site inspection report.

10   **Appointment of Sub-Contractors**

You should have arrangements for appointing competent sub-contractors/consultants.

What criteria is used to assess the competence of any sub-contractors?

11   **Monitoring Performance of Sub-Contractors**

You should have arrangement in place for monitoring sub-contractor performance.

How do you monitor sub-contractor performance?

Section 6 –REFERENCES

12 *You should give details of relevant experience in the field of work for which you are applying.*

Give the names and addresses of 2 organisations where you have completed similar work ideally within the last 12 months.

Provide details of an individual who has first-hand knowledge of your involvement in the project below:

Name:

Address:

Tel No.:   E-mail:

Where there are significant shortfalls in your experience, or there are risks associated with the project that you have not managed previously, you should provide an explanation of how you would overcome these issues:

	Attached(please tick)	
APPENDIX – Evidence to Support Submission	Yes	No

Employer's Liability Insurance

If No please state why?

....

Public Liability (3rd party) Insurance

If No please state why?

....

Health and Safety Policy

If No please state why?

....

Sample Risk Assessments (for similar projects)

If No please state why?

....

Sample Method Statements (for similar projects)

If No please state why?

....

Health and Safety Training Records

If No please state why?

....

Qualifications and Training Certificates

If No please state why?

....

Example of actions taken following an investigation

If No please state why?

....

Example of a site inspection report (from similar project)

If No please state why?

....

Appendix G

Example of a Project Execution Plan (PEP)

SCHEDULE OF REVISIONS

Revision	Date of Revision	Details of Revision	Revised By

Contents

SECTION A – PROJECT

A1 – Project Summary and Key Information
A2 – Regulatory Compliance

SECTION B – PEOPLE

B1 – Project Directory
B2 – Project Structure
B3 – Consultant Schedules of Services
B4 – Contractors' Appointments

SECTION C – PROCESSES

C1 – Communication
C2 – Change Management
C3 – Document Management
C4 – Health and Safety
C5 – VRM

Professional Ethics in Construction and Engineering, First Edition. Jason Challender.
© 2022 John Wiley & Sons Ltd. Published 2022 by John Wiley & Sons Ltd.

APPENDICES

Appendix A – Handover Checklist
Appendix B – Responsibility Matrix
Appendix C – Risk Register
Appendix D – Programme

Section A – Project

A1 – Project Summary and Key Information

1.0   Project Name and Scope

1.1   Project Name

The Project is to be known as '???????', and unless requested by the Project Board, this shall be the only title stated on all documentation including correspondence, meeting notes, documentation and drawing title blocks. Within the body of documents, and in informal communication, the abbreviation '?????' may be used.

1.2   Project Background

Description of why the project is required, and explanation of project to date.

A1 – Project Summary and Key Information

1.3   Project Scope

Outline items to be included within the project.

A1 – Project Summary and Key Information

2.0   Project Plan

2.1   Project Plan

Description of project timescales and the pressure on this.

The project has been established to achieve the following outline programme milestones:

Please give elemental breakdown of programme.

A1 – Project Summary and Key Information

3.0   Project Budget

3.1   Project Budget

Element	Original Cost Plan	Forecast Expenditure	Actual Expenditure	Variance

3.2   Please See Cost Plan in Appendix D

A1 – Project Summary and Key Information

4.0   The Site

4.1   Location

Brief description on location of the building and why this location has been selected

Insert Picture of Location Here

A1 – Project Summary and Key Information

5.0 Abbreviations

5.1 Abbreviations

School of Computing, Science and Engineering	=	CSE
Mechanical and Electrical	=	M & E
Project Manager	=	PM
Post Project Review	=	PPR
Quantity Surveyor	=	QS
Senior Management Team	=	SMT
Construction (Design and Management) Regulations	=	CDM
Principal Designer	=	PD
Change Request Form	=	CRF
Employers Agent (depending on Procurement)	=	EA
Higher Education	=	HE
Programme bar chart (duration, completion)	=	Project Plan

A2 – Regulatory Compliance

1.0 Planning Status

1.1 Planning Consent

Give description on whether planning is required, the type of planning request and likely timescales.

2.0 Building Regulation Status

2.1 Building Regulations

Give description on whether Building Regulations approval is required, the type of planning request and likely timescales.

3.0 Legal Agreements/Landlord/Party Wall/Rights to Light Issues

3.1 Legal Issues

Highlight the potential or foreseen legal issues.

Section B – People

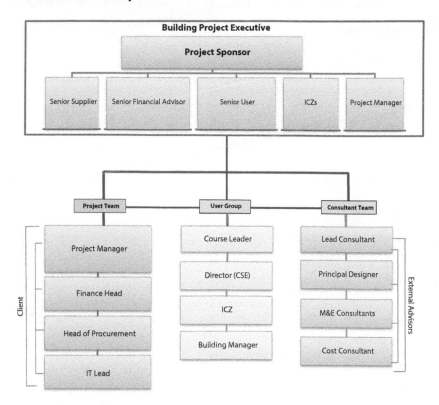

B1 – Project Directory

The Project directory will be maintained and updated by the Project Manager throughout the life of the project.

Role	Company Name/ Address	Contact Name	Phone	Mobile	E-Mail
Project Chair					
Senior User	"				
Senior Supplier	"				
Board Advisor/Head of Financial Accounts TBC	"				
Course Leader	"				
Project Manager	"				

Role	Company Name/Address	Contact Name	Phone	Mobile	E-Mail
Construction/Design Team					
Main Contractor					
Construction Manager					
Lead Consultant					
Principal Designer					
Project Manager					
Lead Designer					

B2 – Project Structure

1.0 Project Structure

1.1 The Combined Project Board/Project Team

The Project Board will encompass the following roles with the following key responsibilities:

- Project Executive – ultimately accountable for the project, key decision maker for project.
- Project Sponsor – responsible for chairing the Project Board and championing the project to internal/external stakeholders.
- Senior User – responsible for clearly defining requirements and coordinating user interests.
- Senior Supplier – accountable for committing or acquiring the resources needed to satisfy the project; has authority to run the project within the constraints laid down by the Project Board.
- Project Assurance – independent monitor of all aspects of the project performance.

The limits of the Senior Supplier and also the Project Board's delegated authority to make decisions are restricted to the works, budget and programme approved within the business case, and only when exceeded should an issue be referred to the Campus Masterplan Programme Board. The Project Board will be accountable for delivering the Project within the delegated authority, and the Board will meet on a needed basis to:

- Accept the schedule of requirements and sign of the final design/plans.
- Respond to any exception reports (where a project cannot proceed within the delegated powers).
- Review and approve project continuation post tender.
- Accept hand over/completion sign off.

B2 – Project Structure

2.0 Meeting Schedule

Initial Project Board Meeting	????
Pre-Start Construction Meeting	????
Mid-Project Review Meeting	????
Handover and Close-Out Meeting	????

The list above is not exhaustive if more meeting are required they will be appropriately arranged.

3.0 Public Relations

2.1 Procedures

The Project Chair is responsible for all external communication relating to the Project, or an appointed designate. Should Project parties wish to communicate externally on project matters they should seek approval from the Project Executive, via the Project Manager before doing so. Should any incidents happen on site, or immediately adjacent, or with regard to the Masterplan development that may be of interest to the general public or news organisations, be it positive or negative, the Project Manager is to be notified immediately to allow the correct course of action to be agreed.

All Programme/Project team members are asked to promote good neighbourly and public relations. The Main Contractor will be required to sign up to, and promote the principles of the Considerate Constructors.

B3 – Consultant Schedules of Services

1.0 Consultant Services
The following consultants have been appointed as follows: – (for contact details please see the Project Directory)

Client -
Principal Designer -
Lead Consultant -
Quantity Surveyor -

All design, specification and cost management services are to be complete by *(enter name)*.

2.0 Responsibility Matrix

Individual scopes of service are contained in the appointment documentation of the relevant consultant. The Responsibility Matrix in Appendix B below details the broad division of responsibilities on the Project.

B4 – Contractors' Appointment

1.0 Main Contractor's Appointment

Please state who the main contractor is and how they have been appointed, include procurement method and evaluation.

B4 – Direct Appointments

1.0 UoS Direct Appointments

Please name any direct appointments made by the client.

Section C – Process

C1 – Communication

1.0 Format/Pro Forma

1.1 Correspondence/Meetings

In order for works to be carried out in a timely and cost-effective fashion, it is important that all members of all project teams communicate effectively and efficiently. It is envisaged that the majority of issues arising throughout the Project will be able to be coordinated by the Project Manager.

Communication between parties involved in the project should predominantly be by oral (telephone and meetings) and electronic (e-mail) means. It is of vital importance that a record is kept of all decisions made concerning the Project, and, therefore, an e-mail (or letter) or minutes should be sent confirming the content of any decision made. E-mail communication should, where possible, only be used to issue information, a set of instructions or confirm a decision made, and should not be used in a 'conversational' manner to avoid inefficient practices.

Communication should follow the routes indicated and be between the key individuals from a relevant organisation. Those key individuals are then responsible for distributing that communication within their organisation.

Alternative communication routes may be followed in the event of an urgent situation or a party/parties not being available, provided that the course of action taken is in the best interest of the Project as a whole. In such a situation, the party/parties who would in the normal way have been involved, must be kept fully informed.

The Project Manager must be copied in on all relevant communications.

C1 – Communication

The hierarchy of correspondence is as follows:

- Minutes
- E-mail
- Conversation/Spoken

2.0   Meetings Schedule

Meetings form one of the prime opportunities for face-to-face communication and are considered fundamental to the successful execution of the project. Meetings will be organised on a needed basis. An agenda for a meeting should be issued with the minutes from the previous meeting. Generally minutes should be a succinct record of the key points discussed at the meeting, with an emphasis on recording the decisions taken and on future actions required.

3.0   Correspondence Distribution

All correspondence and documentation is to be issued with the Project title clearly referenced.

Minutes and other correspondence relating to meetings should be distributed to all meeting invitees, or as appropriate. Thought should be given to ensuring excess distribution is avoided; however, effort should be made to ensure all relevant parties receive information appropriately.

4.0   Action/Information

Each piece of correspondence should be clearly noted for each addressee as either action or information. This will assist all parties to prioritise issues and manage paperwork.

C1 – Communication

5.0   Period for Reply

All parties are obliged to respond to communications within the requested period. If an extension to this period is required, it must be agreed by all relevant parties.

6.0   Reporting Arrangements

All report problems should be addressed to the Project Manager in a timely manner.

7.0 Contract Administration

Drafting of the Contract and the administration of it is to be completed by ????????

C2 – Change Management

1.0 The Process (To Be Reviewed Subject to Procurement Process)

The principle of change management is to establish an approved baseline of information, and if anyone involved in a Project suggests varying from it, a change would be raised for the team to consider, review and for the Senior Supplier to approve and instruct (if acceptable) or to escalate under an exception if it is outside of the scope of the delegated authority.

Unplanned change is highly disruptive and without the explicit control of change, there is a greater chance that project objectives will not be met.

Leadership from the client and active participation by the full project team are critical to the successful operation of a change management process.

The Change Management Procedure will apply to all change irrespective of the originator – e.g. Client, stakeholders, user groups, project team and other third parties.

The Change Management Procedure will apply to all change irrespective of scope, anticipated cost (subject to delegated authority levels).

The change management procedure will be determined by the Employers Agent and agreed with the UoS prior to implementation.

When assessing changes, the project team may find the following classification helpful in determining which instructions are given priority:

- Corrective action – Errors or omissions in work already completed which will result in project objectives not being met.
- Request for change – A change which results in an alteration to the brief, design, specification or project acceptance criteria.

It is recommended that changes required for corrective action are prioritised so that agreed project objectives are met.

The Change control system has two component parts:

- Stage 1 – Early warning notice issued and risk reduction meeting held.
- Stage 2 – Change Control. The formal process used to monitor change on a project. Change control is described in Section C2-3.0

The Stage 2 Change Control Form is used to communicate a need for change and to analyse its impacts. **Up until approval the responsibility for the management of communication is held by the change originator.**

Use of the Stage 2 Change Control Form supports:

- Alerting all parties to all change – potential and actual, whether considered to have any.
- Implications or not.
- Circulating relevant information to all parties.

- Facilitating the assessment of the implications of change prior to formal client approval and instruction.
- Informing client decision making including identification of the actions needed to accommodate a change.
- Ensuring that the impact assessment and client approval does not delay the project.
- Managing change implementation to control the scope of the change so that consequences are within expected limits.
- Enabling the issue of a formal instruction to the project team once a change has been approved.
- Facilitating the central registration of all change.

C2 – Change Management

The key duties and responsibilities in managing the change control system are as follows:

- Client – Sign-off of the change. Sign-off confirms that the client's recommendation is for the change to be implemented and acknowledges that the proposed change is the correct option.
- Originator – Driving the change assessment process prior to sign-off. The originator will complete and circulate the Stage 2 Form and will track and chase responses from the team to meet the agreed assessment timescale.
- Team members – Contribute to the assessment process as set out in Section 3.0. All members of the team have the opportunity to comment on all changes.
- Project Manager/Project Leader – Lead the assessment of completed Stage 2 Forms and manage the sign-off process.

In the event that a change is not identified by its originator prior to implementation, a Stage 1 Early Warning Form will be issued by the Cost or Project Manager to initiate the formal change process, **even if this is after the change has been incorporated into the design.**

The scope of change instructed must be limited to the work described on the Stage 2 Form. Whilst design development will be permitted following sign-off of the change, any material alteration to the scope of the change will require further approval via a further Stage 2 Form.

2.0 Stage 1 – Early Warning

Early warning is a key aspect of the change management system. The approach to early warning will be informal and the assessment requires minimal technical input. However, early warning must be used consistently if abortive work and possible delay and extra cost are to be avoided.

Early warning provides high-level reporting of the cost implications of brief or design development at any stage of a project. It focuses exclusively on issues of affordability, which may include the cost implications of an extended development or build programme.

C2 – Change Management

The Employers Agent will undertake a proactive role through the early warning system to identify change which is not identified by other members of the project team.

No comment on the broader implications of a change is required under the early warning system.

Use of the early warning system will not result in the issue of an approved Stage 2 change order.

The early warning system operates through the issue of a stage 1 form.

Generally, Stage 1 forms will be issued by the Employers Agent as revised information is received. Other Project Team members are encouraged to request the generation of a Stage 1 form by the Employers Agent where they believe that an aspect of the project may be moving outside of its established budget.

3.0   Stage 2 – Change Control Form

The change control procedure is the mechanism by which change orders are approved and issued for action. **No change can be issued without client approval obtained through the change control procedure**.

The change control procedure applies to all change proposed after Contracts have been executed.

Changes that fall within the scope of the procedure include:

- Changes to the brief.
- Changes to project acceptance criteria.
- Changes to design or specification.
- Changes to working method/sequencing that might have an effect on key project outcomes.
- Changes to previously instructed Variation Orders.

C2 – Change Management

The change control procedure is the mechanism by which change orders are approved and issued for action. **No change can be issued without client approval obtained through the change control procedure**.

The change control procedure applies to all change proposed after Contracts have been executed.

Changes that fall within the scope of the procedure include:

- Changes to the brief.
- Changes to project acceptance criteria.
- Changes to design or specification.
- Changes to working method/sequencing that might have an effect on key project outcomes.
- Changes to previously instructed Variation Orders.

The impact of the change will be assessed against the following criteria.

- Project scope and quality.
- Cost.
- Programme.
- Safety.
- Sustainability/Environment.
- Risk/Opportunity.

The impact of change will be assessed against current baselines for cost, programme and Specification as defined in the Contract.

The assessment of the impact of the change should include consideration of a 'do nothing' option.

C2 – Change Management

A request for a Stage 2 Change Control form is to be raised by any party whenever it is considered that there is, or may be a change.

The stage 2 form will be issued by and channelled through the Project Manager for each party to the change process.

4.2   Change Control Contacts

Organisation	Name	E-mail
Client		
Project Manager		
Principal Designer		

4.3   Change References

Change number suffixes are based on initials of each project party. Due to duplication of initials the following references are to be used on the project.

Organisation	Codings
TBC	
TBC	
TBC	
TBC	
TBC	
TBC	
TBC	
TBC	

C2 – Change Management

4.4 Response Times

Response times relating to priority codes on the Stage 2 forms are as follows:

Priority	Response Time
Low	10 working days
Medium	5 working days
High	48 hours

Priority Response time

4.5 Levels of Delegated Authority for Issuing Stage 2 Change Instructions

Subject to any level of delegated authority being agreed by the client, all change instructions must be assessed using the Change Control procedure prior to issue.

Organisation Name E-mail

C3 – Document Management

1.0 Document Control

Throughout the Construction Period the Project Manager will be responsible for co-ordination of formal submissions and responses between the consultant team and the Client.

All members of the Professional Team will maintain sufficient hard and soft copy filing systems to enable effective storage of information originating from that organisation.

Design Information control will be administered by the Contractor. The Project Manager will be responsible for ensuring that all Project Team members comply with the protocols established as part of the system.

C4 – Health and Safety

1.0 Health and Safety

UoS require that the correct attention be given to all Health and Safety matters arising from the Project in accordance with the CDM 2015 Regulations. The client will appoint a Principal Designer who will take the lead in planning, managing, monitoring and co-ordinating health and safety during the project.

All Project Team members have a duty to raise any concerns that they have regarding Health and Safety on the project as soon as they become apparent either directly with the Contractor (when appointed) or the client.

C5 – Value and Risk Management

1.0 Value Management

All members of the team must adopt a value-based approach. This approach involves employing lateral thinking and keeping an open mind to question why things are done in a certain way, and trying to identify more effective and efficient methods (particularly their tasks on the project).

2.0 Risk Management

Risk Management has been established at Project level, and encompasses the review of strategic/high-level risk. Key Project risks are identified to ensure that they are dealt with in a timely and appropriate manner. The Risk Management Strategy for this Programme will differentiate between project risks and project issues as follows:

Risk – A risk is an uncertainty, which, if it occurs, will have a negative impact on the project.

Issue – An issue is something which is certain to happen, or has already happened, which will have a negative impact on the project.

The Risk Management Strategy is, firstly, to identify risks and issues which could negatively affect the chances of achieving the Project objectives; secondly, to determine the perceived relative importance of each risk identified in order to prioritise them with respect to each other; and, thirdly, to discuss how each risk/issue can be mitigated and who should take responsibility in each case. It is vital that all parties actively participate in this process as risk is subjective, and certain risks may be more important to one party than to another. The outputs from this process will be included within the Risk Register, by the PM, following each meeting. A Risk Management Workshop may be held to expand on the above, if deemed necessary. The Risk Management process will review the value criteria established, and focus on the risks to delivering those objectives.

Appendix A

{The following provides a list of typical requirements at completion. The Client requirements and contract documents should be referred to in creating a project-specific list.}

Completion Checklist

AAVT Temporary

		By Whom?	By When?
1.	Practical Completion certificate issued.		
2.	Building Control Certificates of Completion received.		
	Building Control approval confirmed and certificate received.		
	Copy of original outline and detailed approvals issued to Client.		
	Statutory signage installed.		
	Fire-fighting appliances installed and ready for use.		
3.	Design team confirming that Contract Requirements have been met.		
	Consultants confirmation that the Contractors design and the works comply with the Contract Requirements.		
4.	All test certification has been signed by required parties and issued.		
	A copy of all certificates is included in the Building Manual. Test certificates to be included but not limited to:		
	Building Control Completion Certificate		
	17th Edition Electrical Test Certificate		
	Earth Test Certificate		
	Water distribution pipework pressure test certificate		
	Water by-laws compliance certificate		
	Certificate confirming chlorination of water distribution pipe work		
5.	Building Manual, including as-built drawings is complete, approved and issued in required format.		
	Approved As-Built drawings issued.		
6.	Final account agreed.		
	All contract instructions signed off and copies issued.		
7.	Condition surveys agreed between Contractor and Building Management.		
	Building accepted by Building Management and handover recorded.		
8.	All snags rectified and signed off. Procedure agreed for defects and outstanding works.		
	Agreed procedure and programme for the recording and rectification of works or defects outstanding or arising.		

Completion Checklist

AAVT Temporary

		By Whom?	By When?
9.	Statutory and main services authorities satisfied.		
10.	Project Review carried out and attended by all.		

Appendix B

Responsibility Matrix
Project Board/Project Team
The Project Board is ultimately responsible for the project, supported by the Senior User and Senior Supplier. The Board's role is to ensure that the project is focused throughout its life cycle on achieving its objectives and delivering a product that will achieve the forecast benefits.

Throughout the project, the Project Board 'owns' the Business Case.

<u>**Specific Responsibilities**</u>

- Oversee the development of the Project Brief and Business Case.
- Ensure that there is a coherent project organisation structure and logical set of plans.
- Monitor and control the progress of the project at a strategic level, in particular reviewing the Business Case continually (for example, at each end-stage assessment).
- Ensure that any proposed changes of scope, cost or timescale are checked against their possible effects on the Business Case.
- Ensure that risks are being tracked and mitigated as effectively as possible.
- Brief corporate or programme management about project progress.
- Organise and chair Project Board meetings.
- Approve the End Project Report and Lessons Learned Report and ensure that any outstanding Project Issues are documented and passed on to the appropriate body.
- Approve the sending of the project closure notification to corporate or programme management.
- Ensure that the benefits have been realised by holding a post-project review and forward the results of the review to the appropriate stakeholders.

The Project Board is responsible for the overall business assurance of the project – that is, that it remains on target to deliver products that will achieve the expected business benefits, and that the project will be completed within its agreed tolerances for budget and schedule.

Business assurance covers:

- Validation and monitoring of the Business Case against external events and against project progress.
- Keeping the project in line with customer strategies
- Monitoring project finance on behalf of the customer.
- Monitoring the business risks.
- Monitoring any supplier and contractor payments.
- Monitoring changes to the Project Plan to see whether there is any impact on the needs of the business or the project Business Case.
- Assessing the impact of potential changes on the Business Case and Project Plan.
- Constraining user and supplier excesses.
- Informing the project team of any changes caused by a programme of which the project is part (this responsibility may be transferred if there is other programme representation on the project management team).

Senior User

The Senior User is responsible for specifying the needs of those who will use the final product(s), for user liaison with the project team and for monitoring that the solution will meet those needs within the constraints of the Business Case in terms of quality, functionality and ease of use.

The role represents the interest of all those who will use the final product (s) of the project, those for whom the product will achieve an objective or those who will use the product to deliver benefits. The Senior User role commits user resources and monitors products against requirements. This role may require more than one person to cover all the user interests. For the sake of effectiveness the role should not be split between too many people.

Specific responsibilities

- Ensure the desired outcome of the project is specified.
- Make sure that progress towards the outcome required by the users remains consistent from the user perspective.
- Promote and maintain focus on the desired project outcome.
- Ensure that any user resources required for the project are made available.
- Approve Product Descriptions for those products that act as inputs or outputs (interim or final) from the supplier function or will affect them directly.
- Ensure that the products are signed off once completed.
- Resolve user requirements and priority conflicts.
- Provide the user view on Follow-on-Action Recommendations.
- Brief and advise user management on all matters concerning the project.

The assurance responsibilities of the Senior User are to check that:

- Specification of the user's needs is accurate, complete and unambiguous.
- Development of the solution at all stages is monitored to ensure that it will meet the user's needs and is progressing towards that target.

- Impact of potential changes is evaluated from the user point of view.
- Risks to the users are frequently monitored.
- Quality checking of the product at all stages has the appropriate user representation.
- Quality control procedures are used correctly to ensure products meet user requirements.
- User liaison is functioning effectively.

Senior Supplier

The Senior Supplier represents the interests of those designing, developing, facilitating, procuring, implementing and possibly operating and maintaining the project products. This role is accountable for the quality of the products delivered by the supplier(s). The Senior Supplier role must have the authority to commit or acquire supplier resources required.

<u>**Specific Responsibilities**</u>

- Make sure that progress towards the outcome remains consistent from the supplier perspective.
- Promote and maintain focus on the desired project outcome from the point of view of supplier management.
- Ensure that the supplier resources required for the project are made available.
- Approve Product Descriptions for supplier products.
- Contribute supplier opinions on Project Board decisions on whether to implement recommendations on proposed changes.
- Resolve supplier requirements and priority conflicts.
- Arbitrate on, and ensure resolution of, any supplier priority or resource conflicts.
- Brief non-technical management on supplier aspects of the project.
- Attendance at monthly Joint Contractors meeting, the monthly Advice meeting, the Project Board and site visits.
- Sign off samples with support from the technical advisor.
- Line manage the Project Support Office.

The Senior Supplier is responsible for the specialist integrity of the project. The supplier assurance role responsibilities are to:

- Monitor potential changes and their impact on the correctness, completeness and integrity of products against their Product Description from a supplier perspective.
- Monitor any risks in the production aspects of the project.
- Ensure quality-control procedures are used correctly, so that products adhere to requirements.

Project Manager

The Project Manager has the authority to run the project on a day-to-day basis on behalf of the Project Board within the constraints laid down by the board.

The Project Manager's prime responsibility is to ensure that the project produces the required products to the required standard of quality and within the specified constraints of time and cost. The Project Manager is also responsible for the project producing a result capable of achieving the benefits defined in the Business Case.

Specific Responsibilities

- Direct and motivate the project team.
- Plan and monitor the project, and prepare and report to the Project Board through Highlight Reports and End-Stage Reports.
- Take responsibility for overall progress and use of resources and initiate corrective action where necessary.
- Be responsible for change control and any required configuration management.
- Liaise with the Senior Supplier to assure the overall direction and integrity of the project.
- Identify and obtain any support and advice required for the management, planning and control of the project.
- Liaise with any suppliers or account managers.
- Ensure all identified risks are entered into the risk log.
- Day-to-day main point of contact with construction-related activities.
- Attendance at Project Team monthly meeting.
- Work with Technical Advisor to agree sign off for change/detailed design approval.
- Appointment of Technical Advisors, i.e. approval of professional fees.
- Working with the Employers Agent to ensure delivery and administration of the roles in accordance with the contract.
- Ensure coordination of the construction works with the operational activities of the campus.
- Monitoring and reporting to Project Board the overall project budget.

Stakeholder Lead

Specific responsibilities

- Lead on stakeholder engagement with staff and students.
- Provide support to the Project Manager.
- As Head of Facilities ensure via Building Managers that the daily operations of the estates are not impacted by the on-going construction works.

Project Assurance

Assurance covers all interests of a project, including business, user and supplier.

Specific responsibilities

- Thorough liaison between the supplier and the customer is maintained throughout the project.
- User needs and expectations are being met or managed.
- Risks are being controlled.

- The Business Case is being adhered to.
- The right people are planned to be involved in quality checking at the correct points in the product's development.
- Staff are properly trained in the quality-checking procedures.
- The right people are being involved in quality checking.
- The project remains viable.
- The scope of the project is not 'creeping upwards' unnoticed.
- Focus on the business need is maintained.
- Internal and external communications are working.
- Applicable standards are being used.
- Quality assurance standards are being adhered to.

Project Administrative Support

<u>**Specific responsibilities**</u>

The following is a suggested list of tasks:

- Administer change controls.
- Set up and maintain project files.
- Establish document control procedures.
- Collect actuals data and forecasts.
- Administer Project Board meetings.
- Assist with the compilation of reports.

Appendix H

Example of a Health and Safety Contractor's Handbook

It is a condition of the employment by:
The client wishes to make contractors or sub-contractors, understand and take all necessary steps to ensure compliance with their Health and Safety obligations.   In order to assist contractors and sub-contractors to meet their obligations while working on the Estate, the Client has prepared Health and Safety Rules which are required to be read and complied with.   A failure by any contractor to observe the provisions of rules may be viewed by the client's Appointed Officer as a potential breach of these rules. As a minimum, in the event of such a failure, the works will be suspended until the outcome of an initial investigation is known and conditions which are safe and without risk to health are provided. Furthermore, a formal written non-compliance will be produced and a copy registered on the contractor's file for future consideration.   While every effort has been made to cover all important matters, it is not possible to give information covering every possible hazard. Should you wish for any further information or advice please do not hesitate to contact your client Appointed Officer.   No work may take place on the Estate without written authorisation from your client Appointed Officer.   The procedure for obtaining authorisation will be explained during the induction process.

Health and Safety Rules for Contractors

1. CONTACT

   The (client organisation) will appoint an Appointed Officer at the start of a Project. The Contractor will be notified of the name of the Appointed Officer throughout the tender process and related correspondence.

2. REPORTING ATTENDANCE on SITE

   You must initially report to the Estates Helpdesk, giving the name of your Appointed Officer.
   The Appointed Officer will check if they have had an induction in the last 12 months and arrange if required.

Professional Ethics in Construction and Engineering, First Edition. Jason Challender.
© 2022 John Wiley & Sons Ltd. Published 2022 by John Wiley & Sons Ltd.

Out of Hours Access – Following prior approval from your Appointed Officer, report your attendance by signing in and out of the register in Security.

3. <u>PARKING ARRANGEMENTS</u>

Parking spaces are limited.

Parking will only be permitted in designated parking areas as agreed with your Appointed Officer.

Any unauthorised parking, will result in a fine.

Vehicles must not cause any obstruction which would interfere with the normal working of the Company or access by Emergency Services.

Drivers are required to exercise due care and regard for the safety of others.

4. <u>FIRE ALARM and EVACUATION</u>

The fire alarms are tested weekly and will sound for no more than 20 seconds. The Appointed Officer will provide you with the relevant testing times.

Should the Fire Alarm sound for longer or at any other time vacate the building via the nearest exit and go to the Assembly Point which is displayed on notices throughout the buildings.

If you discover a fire you should activate the nearest fire call point, proceed immediately to the Assembly Point and contact Security from a mobile.

If you think you've activated the alarm due to your work activities notify the Building Controller without delay.

DO NOT return to the building until you have been informed it is safe to do so by the Building Controller.

5. <u>BASIC SAFETY RULES</u>

- <u>Smoking</u>

  Smoking is not permitted in any building or near main entrances to buildings. Electronic Cigarettes can be used in offices/rooms which are not inhabited or on view to other staff or students; they must not be used in any circulation spaces/public areas.*

  * Within a site that's been handed over to the Principal Contractor, their rules will apply.

- <u>Refreshment and Toilet Facilities</u>

  If you are permitted to use the Client's refreshment and toilet facilities you should first change out of dirty work wear and keep the facilities clean and tidy. For your own safety please adopt good hand hygiene practices; remember to wash your hands before preparing or eating food.

- <u>Drugs and Alcohol</u>

  No intoxicating liquor or drugs will be allowed on the premises except for Prescription Drugs.

  Where your site workers are required to take prescription drugs, it is your responsibility to ensure their work performance will not be adversely affected whilst on site.

- <u>Conduct</u>

  Ensure that your general behaviour and actions and that of your site workers does not cause any offence or disturbance to any member of the Client Community.

- Guarding
  Never operate with safeguards that have been altered, bypassed or removed.
- Noise
  The use of radios on campus is forbidden, any noise-generating activities including drilling must be pre-arranged with your Appointed Officer to minimise disruption within occupied areas.
- Slips, Trips and Falls
  You must protect others in close proximity to your work area from falling objects, slips, trips and falls and any other risks you may create.
- Speed Limits
  The speed limit on the inner campus is strictly limited to 5 mph.
  The speed limit on Client Road is 10 mph and is subject to all regulations of the Highway Code.
- Housekeeping
  High standards of housekeeping will be maintained at all times and general work areas should be kept clean and free from obstructions.
  Adjacent areas to your work area must be cleaned regularly to avoid any buildup of dust.

6. FIRST AID

To summon First Aid or the Emergency Services you should:
Ring from a mobile, stating the location and nature of the injury.
This is the direct emergency number for the Security Office in Maxwell, please add this number to your mobile, just in case.

7. REPORTING INCIDENTS

All near misses, accidents and incidents must be recorded and reported to the Appointed Officer as soon as reasonably practicable.
This is in addition to any reporting you may do for your own organisation or to the HSE.
You must inform your Appointed Officer immediately if you:
- have, or discover a spillage;
- discover, disturb or damage suspected Asbestos Containing Material
  – STOP THE WORK, seal the area and report it without delay.

8. RISK ASSESSMENTS and WORK PERMITS

Where appropriate, site-specific Risk Assessments and Method Statements (RAMS) must be provided for review by the Appointed Officer in advance of any planned work.
The responsibility for ensuring a safe method of work is adopted rests with the contractor.
Your Appointed Officer will inform you of the hazards; Asbestos, biological, chemical, electrical, mechanical etc. in the area you are working* and any procedures you need to be aware of.
*** Asbestos registers for the areas where you are working will be made available by the Appointed Officer.**
You must obtain Authorisation, in WRITING, before every project is started. This will be in the form of a 'Work Authorisation Certificate' or 'Permit to Work'.

At the end of the planned work the authorisation documents must be signed off and returned to the Appointed Officer.

To be issued with a Work Authorisation Certificate in addition to providing a Risk Assessment and Method Statement (RAMS) for review by the Appointed Officer, the person doing the work must be able to control the hazards, e.g.

- No isolations or line breaking
- No confined space entry
- No excavations
- No hot work
- Access/egress is reasonable

Where the risks can't be controlled via the Work Authorisation Certificate and RAMS alone a Permit to Work may also be required. A Permit to Work is a formalised document which authorises:

- certain people to carry out;
- specific work at a;
- specific site at a;
- certain time and sets out the main precautions needed to complete the job safely.

A hard copy of the permit to work should be clearly displayed at the work site for the duration of the works.

Permits must be signed off and returned to the Appointed Officer at the end of the work.

Where the work will be conducted in a site which has been handed over to the Principal Contractor
then it will be their responsibility to arrange for adequate controls to be put in place.

Typically tasks which may need to be authorised by a Permit to Work will include:

- Asbestos Removal
- Confined Space Entry
- Excavations
- Higher-risk electrical work
- Hot work
- Work on Fire Alarm Systems and Emergency Lighting
- Loft/Ceiling void access
- Roof access/roof work

Contractors should confirm with the Appointed Officer whether or not a Permit to Work is required – this MUST be done prior to commencing the work.

9. ELECTRICAL WORK

You must obtain a Permit to Work from your Appointed Officer prior to:

- Connecting to, or interfering with any electrical or other services.
- Entering any sub-station, switch room or similar area.
- Working on Live Electrical Systems is generally not permitted; except where it is necessary due to the nature of the work, e.g. testing, and a Safe System of Work must be in place.

Any work on electrical systems, however minor, may only be completed by a suitably trained and experienced electrician.

10. ROOF WORK

If it can't be avoided, all work at height must have an appropriate risk assessment and method statement and if your work requires access to a roof your Appointed Officer will provide you with a copy of the relevant roof hazard sheet.

In addition, all roof access must have a roof access permit to work.

11. FIRE SAFETY

All Client buildings are equipped with automatic fire detection systems.

If you are doing anything that might compromise the fire system, for example, generation of dust, blocking fire exits or extinguishers at least 10 days' notice is required to enable the Appointed Officer to make the arrangements to ensure unwanted alarms are prevented and necessary precautions are put in place.

Without exception:

- Flammable materials must be stored securely and appropriately and not left out unattended, particularly at night.
- Dust levels must be kept to a minimum.
- Don't allow combustible materials and debris to accumulate.
- Don't store materials and equipment on stairways and other escape routes.

12. HAZARDOUS SUBSTANCES

All work which is carried out on Site must comply with COSHH Regulations;

- Chemical Substances:
  Prior to bringing chemicals into the Client, e.g. acids and oils, you must provide the Appointed Officer with associated Risk and COSHH assessments for review, and they must be displayed at the site of the works for the duration of the works.
  Your COSHH assessments whilst based on any Safety Data Sheet provided by the manufacturers must also include how the materials will be handled, used, stored, transported and disposed of whilst on site.
- Dust:
  Where equipment is used which is known to generate dust, provision must be made by the contractor to contain the dust, and arrangements must be made to ensure the work is properly supervised.

13. LABORATORY ACCESS

The Client has Biology, Chemistry and Physics Laboratories and numerous Engineering Workshops.

Hazards in these areas may include;

- Harmful organisms
- Hazardous chemicals
- Lasers and Power tools

If you need to access any of these areas:

In addition to your Work Authorisation Certificate, the Appointed Officer will arrange for

The Area Supervisor to provide a 'Certificate of Clearance' identifying any remaining hazards and associated controls.

This will be displayed at the entrance to the room and you need to make sure you are aware of the residual hazards identified.

f there's no certificate on the door or you find anything you are unsure of, STOP 'HE

WORK and report your concerns to the Appointed Officer without delay.

14. <u>LONE WORKING</u>

Anyone working alone must not be placed at any greater risk than any other employee working with others.

Where, following your lone worker risk assessment, there would be additional risk for someone working alone the Client will expect you to provide a second person to be in attendance at all times, so that if anything should happen they can provide or call for assistance.

This is particularly important where your team will be expected to work in high-risk areas such as:

- working at height,
- confined spaces,
- laboratories,
- electrical works and
- work close to exposed live conductors.

15. <u>LIFTING OPERATIONS and EQUIPMENT</u>

All work on site must comply with LOLER Regulations:

Any lifting equipment brought on to the premises must have a copy of its current inspection Certificate, for presentation to the Appointed Officer, before it is used.

Any person using a MEWP shall be adequately trained and hold a current certificate.

UoS cranes, hoists and lifting equipment must not be used by contractors.

16. <u>PERSONAL PROTECTIVE EQUIPMENT</u>

Contractors must provide all appropriate PPE as indicated in the RAMS.

Protective equipment must be used at all times where necessary, regardless of your own views on risk.

Protective equipment must be worn in designated areas including labs.

17. <u>PLANT and EQUIPMENT</u>

Plant, tools, tackle and equipment brought onto site must be fit for purpose, tested, maintained and in good working condition.

Electrical equipment must comply with all current Electricity Regulations and must:

- NOT exceed 110 V without prior permission from the Appointed Officer
- be 'PAT' tested

All machinery brought onto site must comply with the PUWER Regulations (Provision and Use of Work Equipment) and be guarded or fenced appropriately.

At the end of each day you must ensure all your equipment is fully isolated and locked away.

18. <u>VEHICLES on CAMPUS</u>

You must organise your work to allow pedestrians and vehicles to move without risks to health:

Traffic routes should be indicated by warning signs and barriers.

Delivery vehicles must not impede access for emergency vehicles and will only be allowed on site for the loading or unloading to be completed. Deliveries should be pre-planned with someone available to receive the goods or they will be turned away.

Reversing should be kept to a minimum but where required, reversing aids and banksmen should be used.

Extreme caution should be taken whilst driving or operating machinery on campus due to the large numbers of students moving between buildings (particularly at the start and end of lectures). It is essential that pedestrians and vehicles are segregated.

19. WASTE MANAGEMENT

The Contractor is responsible for the removal of all waste from site in accordance with current environmental legislation.

Unless specifically authorised you must not place debris into skips controlled by the Client of Salford.

Skips left on site MUST be of the self-contained lockable type and their location agreed by the Appointed Officer.

Care must be taken not to discharge trade effluent or contaminated liquids into the drainage system or water courses, e.g. adequate storage facilities must be provided for diesel fuel to ensure containment and prevent spillage.

Detailed records must be kept of all waste removed from site including the type and volume of waste removed from site and the method of disposal (landfill or recycled).

20. WORK at HEIGHT

All work at height must comply with Work at Height Regulations; scaffolds, ladders and other access equipment must be in sound condition and of good construction, adequate for the purpose and properly maintained.

If as a result of a Risk Assessment ladders are identified as an appropriate control, as a minimum they must be:

- used for access and egress only or;
- for work of short duration that is considered to be low risk.

Ladders must be inspected before use to ensure they are in a safe condition and they must be secured adequately before use.

Ladders must be taken down after use or at the end of the day.

Unattended ladders and ropes must be secured out of reach of students and other unauthorised persons.

UoS equipment, including ladders must not be used by Contractors.

21. CONTRACTOR in CONTROL

When a site is handed over to the Contractor (including the Principal Contractor), as a minimum we expect that:

A suitable site induction is provided to all construction site workers taking into account, but not limited to:

- the information included in this induction;

- any site specific risks and control measures that those working on the project need to know about;
- first aid arrangements;
- accident and incident reporting arrangements.

Necessary steps are taken to prevent access by unauthorised persons to the construction site including:

- Physically defining the site boundaries using suitable barriers and warning signs.
- Special consideration of the nature of the business is given (adjoining areas with student/staff access).
- Changing fence lines and access routes can only be carried out in agreement with the Appointed Officer.

Provision of suitable and sufficient welfare facilities.

The Client reserves the rights to carry out periodic site inspections to assess compliance with control measures.

FINAL NOTE:

It is your responsibility to ensure that all the information provided in this document, that is relevant to your works, is included in your risk assessments and method statements both for dealing with the issues raised as well as the work you will be doing. You will then need to ensure that all the site and safety information is passed on to your staff, sub-contractors and anyone else that comes onto your site during the contract, through site inductions, tool box talks and any other means appropriate.

Appendix I

RICS Regulations and Guidance Notes

9.1   The Global Professional and Ethical Standards

There are five standards. All members must be able to demonstrate that they:

Act with integrity
- This means being honest and straightforward in all that you do.
- This standard includes, but is not limited to the following behaviours or actions:
  - Being trustworthy in all that you do.
  - Being open and transparent in the way you work. Sharing appropriate and necessary information with your clients and/or others to conduct business and doing so in a way so they can understand that information.
  - Respecting confidential information of your clients and potential clients. Don't divulge information to others unless it is appropriate to do so.
  - Not taking advantage of a client, a colleague, a third party or anyone to whom you owe a duty of care.
  - Not allowing bias, conflict of interest or the undue influence of others to override your professional or business judgements and obligations.
  - Making clear to all interested parties where a conflict of interest, or even a potential conflict of interest, arises between you or your employer and your client.
  - Not offering or accepting gifts, hospitality or services, which might suggest an improper obligation.
  - Acting consistently in the public interest when it comes to making decisions or providing advice.
- Some of the key questions that you could ask yourself include: o What would an independent person think of my actions?

  - Would I be happy to read about my actions in the press?
  - How would my actions look to RICS?
  - How would my actions look to my peers?
- Do people trust me? If not, why not?

  - How often do I question what I do, not just in relation to meeting technical requirements but also in terms of acting professionally and ethically?
  - Is this in the interest of my client, or my interest, or the interest of someone else?

Professional Ethics in Construction and Engineering, First Edition. Jason Challender.
© 2022 John Wiley & Sons Ltd. Published 2022 by John Wiley & Sons Ltd.

- Would I like to be treated in this way if I were a client?
- Do I promote professional and ethical standards in all that I do?
- Do I say "show me where it says I can't" or do I say "is this ethical"?

Always Provide a High Standard of Service

- This means always ensuring that your client, or others to whom you have professional responsibility, receive the best possible advice, support or performance of the terms of engagement you have agreed.
- This standard includes, but is not limited to, the following behaviours or actions:

  - Be clear about what service your client wants and the service you are providing.
  - Act within your scope of competence. If it appears that services are required outside that scope then be prepared to do something about it, or example, make it known to your client, obtain expert input or consultation, or if it is the case that you are unable to meet the service requirements, explain that you are not best placed to act for the client.
  - Be transparent about fees and any other costs or payments such as referral fees or commissions.
  - Communicate with your client in a way that will allow them to make informed decisions.
  - If you use the services of others then ensure that you pay for those services within the timescale agreed.
  - Encourage your firm or organisation you work for to put the fair treatment of clients at the centre of its business culture.
- Some of the key questions that you could ask yourself include: o Do I explain clearly what I promise to do and do I keep to that promise?
  - Do I look at ways to improve the service I provide to my clients?
  - How can I help my clients better understand the surveying services that I am offering?
  - Am I providing a professional service for a professional fee?
  - Would the client still employ me if they knew more about me and the workload I have? If not, why not?
  - Do I put undue pressure on myself and colleagues (especially junior colleagues) to do more than we actually can?

Act in a Way that Promotes Trust in the Profession

- This means acting in a manner, both in your professional life and private life, to promote you, your firm or organisation you work for and the profession in a professional and positive way.
- This standard includes, but is not limited to, the following behaviours or actions:
  - Promoting what you and the profession stand for – the highest standards globally.
  - Understanding that being a professional is more than just about how you behave at work; it is also, about how you behave in your private life.

- Understanding how your actions affect others and the environment and if appropriate questioning or amending that behaviour.
- Fulfilling your obligations. Doing what you say you will.
- Always trying to meet the spirit of your professional standards and not just the letter of the standards.
- Some of the key questions that you could ask yourself include:

- Do my actions promote the profession in the best light possible?
- What is the best way for me to promote trust in me, my firm and the profession?
- Do I explain and promote the benefits, the checks and balances that exist with the professional services that I provide?

Treat Others with Respect

- This means treating people with courtesy, politeness and consideration, no matter their race, religion, size, age, country of origin, gender, sexual orientation or disability. It also means being aware of cultural sensitivities and business practices.
- This standard includes, but is not limited to, the following behaviours or actions:
  - Always being courteous, polite and considerate to clients, potential clients and everyone else you come into contact with.
  - Never discriminate against anyone for whatever reason. Always ensure that issues of race, gender, sexual orientation, age, size, religion, country of origin or disability have no place in the way you deal with other people or do business.
  - As much as you are able, encourage the firm or organisation you work for to put the fair and respectful treatment of clients at the centre of its business culture.
- Some of the key questions that you could ask yourself include:

- Would I allow my behaviour or the way I make my decisions to be publicly scrutinised? If not, why not? If so, what would the public think?
- Are my personal feelings, views, prejudices or preferences influencing my business decisions?
- How would I feel if somebody treated me this way?
- Do I treat each person as an individual?

Take Responsibility

- This means being accountable for all your actions – don't blame others if things go wrong, and if you suspect something isn't right be prepared to do something.
- This standard includes, but is not limited to, the following behaviours or actions:
  - Always act with skill, care and diligence.
  - If someone makes a complaint about something that you have done then respond in an appropriate and professional manner and aim to resolve the matter to the satisfaction of the complainant as far as you can.
  - If you think something is not right, be prepared to question it and raise the matter as appropriate with your colleagues, within your firm or the organisation that you work for, with RICS or with any other appropriate person, body or organisation.

- Some of the key questions that you could ask yourself include: o Am I approachable?

  - Does my firm or organisation have a clear complaints handling procedure?
  - Do I learn from complaints?
  - Do I take complaints seriously?
  - Am I clear about what the process is within my firm or the organisation that I work for about raising concerns?
  - Have I considered asking for advice from RICS?

Appendix J

RICS Frequently Asked Questions Document Linked to Their Global Professional and Ethical Standards

Global Professional and Ethical Standards
Frequently Asked Questions

Q) What are the New Global Professional and Ethical Standards?
A) There are five standards:

- Act with integrity
- Always provide a high standard of service
- Act in a way that promotes trust in the profession
- Treat others with respect
- Take responsibility.

The new global standards aim to provide clarity and simplicity around what is expected from RICS members.

Q) Why are you introducing new professional and ethical standards?

A) RICS' increased global approach to standards and regulation provide the perfect opportunity to look at the ethical framework for the profession and how it will work best on a global scale. We have done so by undertaking extensive consultation globally and talking to experts in the field of ethics. This resulted in RICS Governing Council agreeing new global professional and ethical standards. It also agreed the production of supporting information, advice and guidance to assist members in meeting those standards and to make those difficult ethical decisions.

Q) Why are there now only 5 standards instead of the previous 12 ethical values?

A) The standards were developed after considerable consultation and discussion globally. The new standards cover everything that the previous 12 values covered, but do so in a much clearer way and they have been designed to be applicable globally.

Q) How do the professional and ethical standards link in with the RICS rules of conduct for members and firms?

A) The two really go hand in hand. Both the rules of conduct for members and for firms require respectively members and firms to act with integrity. The global professional

Professional Ethics in Construction and Engineering, First Edition. Jason Challender.
© 2022 John Wiley & Sons Ltd. Published 2022 by John Wiley & Sons Ltd.

and ethical standards go into more detail around what integrity and the other ethical standards mean. Not meeting the requirements of the rules of conduct or the ethical standards, or both, may lead to investigation and possible disciplinary action.

There is an even bigger link here in that all members meet high standards set by the profession, in terms of entry standards, technical standards, regulatory standards and ethical standards. Meeting those standards and enforcing those standards is the professions promise of trust to the public.

Q) Are there any guidance materials to help me learn more about the new standards?

A) Yes. There are supporting levels of information available for each standard, including:

- A brief explanation of what each standard means.
- Examples of the types of actions or behaviours that would demonstrate the standards and questions that members will wish to consider further information around some of the more frequent ethical areas that members may face. Such as conflicts of interest, raising concerns when it may appear standards are not being met or there is inappropriate behaviour, and the interplay between ethics, laws, rules, regulations, and entry, technical and regulatory standards.
- A range of case studies that illustrate each of the standards. These examples are real-case studies we have collected and link to ethical issues that relate to different specialty areas within surveying and different geographical areas.

Q) Will you translate the standards into different languages for different world regions?

A) Yes, the standards will be translated into many languages including Simplified Chinese, French, German, Spanish, Portuguese, Dutch, Polish and Russian.

Q) I always work in an ethical manner; why are the standards important to me?

A) By far the vast majority of members work in an ethical manner. It is only a very small percentage of members that have to go through RICS Regulatory processes. The new standards are not there to criticise members, quite the reverse, they are there to promote how members work and the high standards they meet on a day-to-day basis. With a membership that is approximately 180,000 strong (which includes approximately 80,000 students and trainees) globally it is important to have a set of professional and ethical standards that anchors the whole profession. It is also important to remember that members will be at different stages in their careers, areas of specialism and in different-sized firms. The standards and supporting information are there to help members and provide appropriate guidance and advice.

Q) My firm has its own ethical standards for staff; do I still need to adhere to the RICS standards?

A) Yes, you do. RICS' ethical standards are the standards for the whole profession. You may well be in a position where the firm you work for has ethical standards or values as well. In practice this shouldn't be a problem as it is difficult to see how there can be real conflict between the RICS standards and your firm's

standards. You are likely to find similar words used, for example, integrity and respect for others. If there is a conflict then you should apply whatever is the higher test.

Q) Whilst it is right for RICS members to meet ethical standards, there are others in the market place that do not and aren't we just giving them an advantage here?

A) No. Acting ethically and doing business professionally have been shown to increase business. The majority of clients are more likely to place business with you or your firm. They are also more likely to promote your services to others if they know about the standards you work to and whether you behave in a professional and ethical manner.

Q) I am going through my Assessment of Professional Competence (APC) at the moment and have completed my work so far using the existing 12 ethical principles that were in place before the introduction of the new global and professional ethical standards. How does this affect me and what should I do?

A) Don't worry, there is going to be an orderly transition. Current APC candidates will not be affected by the introduction of the new ethical standards. All candidates must be assessed on ethics and the 'old' standards will be used for session one 2012. From 1 June 2012* candidates preparing for final assessment should refer to the new standards. *European candidates: effective date 1 September 2012.*

Glossary of Terms

Buildability An assessment of design details in respect of construction sequencing and activity.

Collaboration "Working jointly on an activity or project" (Oxford Current English Dictionary, 1990) and collaborative working as "the act of people working together toward common goals" (National B2B Centre, 2007).

Contextualisation To study or consider something in context.

Epistemological Philosophy. The theory of knowledge, particularly in respect of its scope, methods and validity.

Framework "An agreement with suppliers to establish terms governing contracts that may be awarded during the life of an agreement. A general term for agreements that set out terms and conditions for making special purchases" (Constructing Excellence, 2015).

Humanistic Rationalist system of thought or viewpoint attaching prime importance to human rather than divine

Objectivistic Philosophy. The belief that some things, particularly moral truths exist independently of human knowledge.

Partnering "A structured management process to focus the attention of all parties on problem resolution" (Larsen, 1997) or alternatively "business relationships designed to achieve mutual objectives and benefits between contracting organisations" (Wong and Cheung, 2004).

Profit A financial gain, especially between amount earned and the amount spent in buying, operating or producing something (Oxford Dictionaries, 2014).

Profitability Financial measure of success of an organisation through profit.

Scape Public sector owned specialist offering a full suite of OJEU compliant national procurement frameworks.

Subjectivism Philosophy. The doctrine that there is no objective or external truth and that knowledge is merely based or influenced by personal feelings, tastes or opinions.

Trust "Expectancy held by an individual that the word, promise, oral or written statement of another individual or group can be relied upon" (Rotter, 1980) or conversely as a "belief in a person's competency" (Sitkin and Roth). The Oxford Current English Dictionary (1990) offers a further definition as "confidence in or reliance on some quality or attribute of a person or thing or the truth of a statement."

Professional Ethics in Construction and Engineering, First Edition. Jason Challender.
© 2022 John Wiley & Sons Ltd. Published 2022 by John Wiley & Sons Ltd.

Index